The study of the interaction of proteins, whether they
proteins or structural proteins, with nucleic acids is
areas of modern molecular biology. Professor Steitz
research in structural studies of DNA-binding prote
with DNA. The author clearly and concisely describes the use of new techniques in molecular genetics, DNA synthesis, protein crystallography and nuclear magnetic resonance in the field.

The book is essential reading for research workers and students entering this exciting area of research.

Structural studies of protein–nucleic acid interaction: the sources of sequence-specific binding

Structural studies of protein–nucleic acid interaction

the sources of sequence-specific binding

THOMAS A. STEITZ

Department of Molecular Biophysics and Biochemistry
and Howard Hughes Medical Institute at Yale University

Published by the Press Syndicate of the University of Cambridge
The Pitt Building, Trumpington Street, Cambridge CB2 1RP
40 West 20th Street, New York, NY 10011–4211, USA
10 Stamford Road, Oakleigh, Melbourne 3166, Australia

First published 1990 in *Quarterly Reviews of Biophysics*
First published as a book 1993

Printed in Great Britain at the University Press, Cambridge

A catalogue for this book is available from the British Library

Library of Congress cataloguing in publication data

Steitz, Thomas A.
Structural studies of protein–nucleic acid interaction: the sources of sequence-specific binding / Thomas A. Steitz.
p. cm.
Originally published in Quarterly reviews of biophysics in 1990.
Includes bibliographical references.
ISBN 0-521-41489-X (pbk.)
1. DNA-protein interactions. 2. RNA-protein interactions.
I. Title.
QP624.75.P74S84 1993
574.87′328 – dc20 92-41805 CIP

ISBN 0 521 41489 X paperback

UP

Contents

1. INTRODUCTION AND OVERVIEW

Structural studies of DNA-binding proteins and their complexes with DNA have proceeded at an accelerating pace in recent years due to important technical advances in molecular genetics, DNA synthesis, protein crystallography and nuclear magnetic resonance. The last major review on this subject by Pabo & Sauer (1984) summarized the structural and functional studies of the three sequence-specific DNA-binding proteins whose crystal structures were then known, the *E. coli* catabolite gene activator protein (CAP) (McKay & Steitz, 1981; McKay *et al.* 1982; Weber & Steitz, 1987), a *cro* repressor from phage λ (Anderson *et al.* 1981), and the DNA-binding proteolytic fragment of λcI repressor protein (Pabo & Lewis, 1982) Although crystallographic studies of the *E. coli lac* repressor protein were initiated as early as 1971 when it was the only regulatory protein available in sufficient quantities for structural studies (Steitz *et al.* 1974), little was established about the structural aspects of DNA-binding proteins until the structure of CAP was determined in 1980 followed shortly thereafter by the structure of λ*cro* repressor and subsequently that of the λ repressor fragment. There are now determined at high resolution the crystal structures of seven prokaryotic gene regulatory proteins or fragments [CAP, λ*cro*, λcI repressor fragment, 434 repressor fragment (Anderson *et al.* 1987), 434 *cro* repressor (Wolberger *et al.* 1988), *E. coli trp* repressor (Schevitz *et al.* 1985), *E. coli met* repressor (Rafferty *et al.* 1989)], *Eco*R I restriction endonuclease (McClarin *et al.* 1986), DNAse I (Suck & Ofner, 1986), the catalytic domain of $\gamma\delta$ resolvase (Hatfull *et al.* 1989) and two sequence-independent double-stranded DNA-binding proteins [the Klenow fragment of *E. coli* DNA polymerase I (Ollis *et al.* 1985) and the *E. coli* Hu protein (Tanaka *et al.*, 1984)].

Structural studies of DNA-binding proteins complexed with appropriate DNA fragments were slowed initially by the unavailability of large quantities of pure DNA fragments of specific length. The first results of crystallographic studies of the *Eco*R I–DNA complex were reported in 1985 (Frederick *et al.* 1984). High-resolution crystal structures of protein–DNA complexes have now been reported in detail for 434 repressor fragment (Anderson *et al.* 1987; Aggarwal *et al.* 1988), Klenow fragment of *E. coli* DNA polymerase I (Steitz *et al.* 1987; Freemont *et al.* 1988), DNAse I (Suck *et al.* 1988), λcI repressor fragment (Jordan & Pabo, 1988), and *E. coli trp* repressor (Otwinowski *et al.* 1988) and the structures of several more protein complexes with DNA (CAP, *E. coli met* repressor and λ*cro*) have been determined, but are as yet unpublished. There are also the first structural determinations of the DNA binding fragments determined by 2D NMR techniques; these include the DNA-binding fragment of *E. coli lac* repressor (Kaptain *et al.* 1985), *Drosophila* homeodomain of the *Antennapedia* homeotic protein from *Drosophila melanogoster* (Qian *et al.* 1989) and zinc 'finger' domains of eukaryotic transcription factors (Paragga *et al.* 1988; Lee *et al.* 1989). Progress in the determination of the structures of DNA-binding proteins and their complexes with DNA are summarized in Table 1. Equally important to our understanding of the function of these DNA-binding proteins have been the

Table 1. *High-resolution structures of DNA-binding proteins and complexes*

Protein name	Resolution	Function	Reference
E. coli CAP–cAMP	2·9, 2·5 Å	Transcription activator, repressor	McKay & Steitz (1981), Weber & Steitz (1987)
λ*cro*	2·2 Å	Repressor	Anderson *et al.* (1981)
λ*cI* repressor fragment	2·5 Å	Repressor	Pabo & Lewis (1982)
Gene 5 protein from fd phage	2·3 Å	Single-strand DNA binding	Brayer & McPherson (1983)
E. coli Hu	3·0 Å	Chromosome structure	Tanaka *et al.* (1984)
Bovine DNase I	2·5 Å	Duplex DNase	Suck *et al.* (1984)
	2·0 Å		Suck & Oefner (1986)
E. coli Klenow fragment	3·3 Å	DNA polymerase, 3′5′-exonuclease	Ollis *et al.* (1985)
E. coli lac repressor 'headpiece'	NMR	DNA binding	Kaptain *et al.* (1985)
E. coli trp repressor *trp*	2·6 Å	Repressor	Schevitz *et al.* (1986)
trp repressor (*apo*)	1·8 Å	Repressor	Lawson *et al.* (1988)
$\gamma\delta$ resolvase catalytic domain	2·6 Å	Recombinase, repressor	Hatfull *et al.* (1989)
E. coli met repressor	1·7 Å	Repressor	Rafferty *et al.* (1989)
ADR 1 'zinc finger'	NMR	DNA binding	Parraga *et al.* (1988)
Xenopus xfin 'zinc finger'	NMR	DNA binding	Lee *et al.* (1989)
Antennapedia homeodomain	NMR	DNA binding	Qian *et al.* (1989)
E. coli recA protein	2·8 Å	Recombination	Story & Steitz (1990), unpublished
Protein–DNA complexes			
*E. coli Eco*R I–DNA	3·0 Å	Restriction enzyme	McClarin *et al.* (1986)

434 repressor fragment–DNA	4·5–3·5 Å	Repressor	Anderson *et al.* (1987)
	2·5 Å		Aggarwal *et al.* (1988)
E. coli Klenow fragment–DNA	3·3 Å	3′5′-exonuclease	Steitz *et al.* (1987)
E. coli trp repressor–DNA	2·4 Å	Repressor	Otwinowski *et al.* (1988)
λ repressor fragment–DNA	2·6 Å	Repressor	Jordan & Pabo (1988)
Bovine DNase I–DNA	2·0 Å	Duplex DNase	Suck *et al.* (1988)
434 *cro* repressor–DNA	≈ 4 Å	Repressor	Wolberger *et al.* (1988)
E. coli Met repressor–DNA	?	Repressor	Rafferty *et al.* (1989), note added in proof
E. coli CAP–cAMP–DNA	3·0 Å	Transcription activator	Schultz, *et al.* (1989) unpublished

studies emanating from molecular genetic approaches to correlate structure with function.

Less rapid progress has been made on determining the structures of protein complexes with folded RNA molecules and less can be surmised about the nature of protein–RNA interaction from the structures of RNA-binding proteins alone. The structures of *B. stearothermophilus* tyrosyl-tRNA synthetase (tyrRS) (Bhat *et al.* 1982; Brick *et al.* 1989), *E. coli* methionyl-tRNA synthetase (MetRS) (Zelwer *et al.* 1982; Brunie *et al.* 1987) and *E. coli* elongation factor TU (Jurnak, 1985; La Cour *et al.* 1985) have been determined in the absence of tRNA. Recently, however, the structure of *E. coli* glutaminyl-tRNA synthetase (GlnRS) complexed to $tRNA^{Gln}$ and ATP has been determined and provides a wealth of information on the details of protein–RNA interactions (Rould *et al.* 1989; Perona *et al.* 1989).

Major contributors to this remarkable set of scientific advances have been technical advances in molecular genetics, DNA synthesis, increased speed and power of computation, rapid methods of data collection in protein crystallography and the development of 2D NMR methods of structure determination.

Clearly the ability to clone and overexpress the protein and RNA gene products for these normally rare nucleic acid-binding proteins has made accessible a large variety of proteins in this class. Co-crystallization of proteins with the ligands and substrates to which they bind has been a standard technique for examining the source of their specificity in detail over since the determination of the structure of lysozyme complexed with a trisaccharide was established in 1965 (Johnson & Phillips, 1965). Only after the technology of DNA synthesis advanced to the stage that tens of milligrams of a specific-sequence oligonucleotide could be made and purified in the early 1980s was it possible to co-crystallize proteins with DNA and carry out similar studies on DNA-binding proteins. The current availability of DNA-synthesizing machines and high-pressure liquid chromatography makes this technology now accessible to non-organic chemical laboratories. Similarly, the ability now to make large quantities of specific RNA species either by cloning its gene into a high expression vector or by *in vitro* transcription by T7 RNA polymerase makes the study of protein–RNA complexes accessible to crystallographic analysis. The advent of rapid X-ray crystallographic data collection techniques provided by 2D area detectors has only begun to have a dramatic effect on the pace of structural studies of macromolecules.

2. PRINCIPLES OF SEQUENCE-SPECIFIC NUCLEIC-ACID RECOGNITION

Structural, biochemical and molecular-genetic studies of protein–nucleic acid complexes have established two important sources of sequence specificity in protein–nucleic acid interactions: (1) Direct hydrogen bonding and van der Waals interaction between protein side chains and the exposed edges of base pairs, primarily in the major groove of B-DNA and to a lesser extent the minor grooves of DNA and RNA provides structural complementarity to correct but not to incorrect sequences. (2) The sequence-dependent bendability or deformability of duplex DNA or RNA provides sequence selectivity by virtue of the ability of some nucleic-acid sequences to take up a particular structure required for binding to a protein at lower free energy cost than other sequences.

2.1 *The problem that is set: what is being recognized*

We are concerned here primarily with the interaction between proteins and duplex DNA and RNA, often in a nucleic-acid sequence-dependent fashion. The 3D structure of double-stranded DNA is highly polymorphic (Kennard & Hunter, 1989) but variants of two forms, A-form and B-form, are of relevance to the proteins described here. Fig. 1 shows an important difference between A- and B-form DNA. In B-DNA the major groove is wide enough to accommodate an α-helix and the functional groups on the exposed edges of the base pairs can be contacted by protein; the minor groove, on the other hand, is deep and narrow (5·7 Å) and thus less accessible to secondary structures such as an α-helix. For A-DNA (or RNA which is always A-form) the opposite is true. The minor groove is shallow and broad (10–11 Å wide for RNA) whereas the major groove is very deep and narrow [4 Å for RNA (Delarue & Moras, 1989)]. The width of the minor groove in B-DNA varies depending on base composition. AT rich sequences have a narrower minor groove (3·5 Å) than GC rich sequences (Yoon *et al.* 1988). In general, however, one might expect proteins to directly decode DNA sequences via interactions in the major groove but discriminate among RNA sequences via interactions in the minor groove. This appears for the most part to be the case.

A second important consideration in the suitability of the major and minor grooves for direct sequence recognition is the degree of structural variation of the four base pairs as viewed from the two grooves. Seeman *et al.* (1976) pointed out that the base pairs present a more richly varied set of hydrogen bond donors to the major groove as compared to the minor groove. Figure 2 shows that the minor groove side of the base pairs is a veritable recognition desert with only the N2 of guanine distinguishing AT from GC. The patterns of donors and acceptors on the major groove side, however, can distinguish all four base pairs. It is possible, therefore, that direct minor groove recognition can distinguish only a binary code (GC or CG *vs.* AT or TA) whereas major groove recognition can discriminate among all four base pairs.

2.2 *Role of the major groove in DNA recognition*

The extensive hydrogen bonding and shape complementarity between the major groove of B-DNA and the surfaces of many of the sequence-specific DNA-binding proteins has been extensively documented from high-resolution crystal structures of DNA complexes with both the 434 and λ repressor DNA-binding fragments (Aggarwal *et al.* 1988; Jordan & Pabo, 1988), the 434 *cro* repressor (Wolberger *et al.* 1988) and *Eco*R I (McClarin *et al.* 1986). [An interesting variation involving water-mediated hydrogen bonding in the major groove is seen in the *trp* repressor complex with DNA (Otwinowski *et al.* 1988).] In general, structural complementarity between a protein and a specific DNA sequence is achieved in idiosyncratic manners: there does not appear to be a code for nucleic-acid sequence recognition (Pabo, 1983; Matthews, 1988). While particular amino-acid side chains do not always recognize the same base pair, there are some apparent preferences as suggested by Seeman *et al.* (1976). The guanidinium group of arginine often makes a bidentate interaction with the N7 and O6 of guanine (observed with *Eco*R I (McClarin *et al.* 1986) and *Trp* repressor (Otwinowski *et al.*, 1988) and proposed for *λcro* repressor (Ohlendorf, *et al.* 1982) and CAP (Weber and Steitz, 1984). However, it is also observed to interact with the N7s of two adjacent adenines in *Eco*R I (McClarin *et al.*, 1986). Similarly, while the hydrogen-bond donors and acceptors of the glutamine side chain are observed frequently to interact with the corresponding hydrogen-bond donors and acceptors on adenine (Aggarwal *et al.* 1988; Jordan & Pabo, 1988), other interactions of glutamine are also seen, such as the interaction of its NH_2 with the O6 and N7 of guanine in 434 repressor. Perhaps somewhat unexpectedly the carboxylate side chain of glutamic acid is observed in the case of *Eco*R I (Frederick *et al.* 1985) and proposed in the case of CAP (Weber & Steitz, 1984) to simultaneously interact with two adjacent base pairs (the N6s of two adenines in *Eco*R I). The ability of these side chains to make bidentate interactions with DNA greatly enhances their suitability for sequence-specific recognition (Seeman *et al.* 1976). The van der Waals interactions between the protein and the 5-methyl group of thymine appear also to contribute to specificity. Presumably, the close packing of protein against a GC base pair would, in many cases, sterically exclude its replacement by an AT base pair with its accompanying bulky 5-methyl group.

It has frequently been proposed that the sequence specificity of a protein might be altered by changing a side chain such as arginine that recognizes the guanine in a GC base pair for a side chain such as glutamine that might recognize the adenosine in an AT base pair. Close examination of the detailed crystal structures of DNA complexes as well as simple model building suggests that this is not likely to work. First of all, the glutamine and arginine side chains are of different lengths and are thus not strictly interchangeable, other features of protein and nucleic acid being held constant. Secondly, those protein side chains interacting with the bases can be involved in a more extensive network of interactions within the protein allowing the formation of a complex protein surface. Thus, for example, Glutamine 44 in *λ* repressor (Jordon & Pabo, 1988) is interacting both with an adenosine and with glutamine 33 that in turn is also interacting with the backbone phosphate.

2.3 *Role of nucleic acid bendability*

The extent to which the sequence-dependent bendability or deformability of DNA would play a vital role in nucleic-acid sequence recognition was perhaps less well anticipated. It appears, however, to play an extremely important role in most though perhaps not all of the complexes. Richmond & Steitz (1976) concluded that '...sequence-dependent alteration of the double-stranded DNA conformation are probably important in *lac* repressor specificity' in order to account for the two to three orders of magnitude difference in *lac* repressor affinity for poly[d(AUHgX)] with various different bulky substituents at the 5 position of U. Crick & Klug (1975) hypothesized that the wrapping of DNA around the histone core in the nucleosome might be achieved by periodic sharp bends or 'kinks' rather than smooth bending. In spite of a few early hints no direct evidence for the role of DNA distortion was obtained until co-crystal structures of complexes were determined.

Significant distortion of the DNA or RNA structure from its presumed structure in solution is observed in the crystal structures of DNA complexes with *Eco*R I (Frederick *et al.* 1985; McClarin *et al.* 1986), 434 repressor (Aggarwal *et al.* 1988), *trp* repressor (Otwinowski *et al.* 1988), DNAse–I (Suck *et al.* 1988), Klenow fragment (Steitz *et al.* 1987; Freemont *et al.* 1988) and the tRNA complex with GlnRS (Rould *et al.* 1989). The structure of the nucleosome although at low resolution shows significant kinking in the DNA (Richmond *et al.* 1984). Model building (Warwicker *et al.* 1987) and the recent determined co-crystal structure (Schultz, Shields and Steitz, unpublished) as well as DNA binding and gel shift experiments (Wu & Crothers, 1984; Gartenberg & Crothers, 1988) establish an important role for DNA kinking and bendability in the sequence-specific binding of CAP to DNA, as delineated in detail below.

The distortions of duplex DNA structure that have been observed in complexes include changes in twist (Aggarwal *et al.* 1988), groove width (Suck *et al.* 1988) and kinks (McClarin *et al.* 1986; Otwinowski *et al.* 1988; Aggarwal *et al.* 1988; Warwicker *et al.* 1987; Schultz, Shields & Steitz 1990, unpublished). The two types of kinks observed thus far are an abrupt reduction in the twist between adjacent base pairs which widens the major groove (McClarin *et al.* 1986) and a change in the roll-angle between successive base pairs resulting in a bend in the DNA helix axis. As initially described by Drew & Travers (1984) AT-rich sequences favour bending into the minor groove while GC-rich sequences facilitate kinks that narrow the major groove.

The sequence-dependent deformability of duplex DNA or RNA that provides specificity for sequences being recognized by the protein can include the melting of base pairs. If binding to protein requires melting of one or more base pairs, then the binding of mismatched base pairs should be favoured over AT pairs that in turn should bind better than GC pairs. The order of binding should reflect the thermodynamic stability of the base pairs.

Two examples of the role of duplex meltability in sequence specificity can be cited. Binding of $tRNA^{Gln}$ to its cognate synthetase results in the breaking of the base pair between nucleotides U_1 and A_{72} (Rould *et al.* 1989). For GlnRS recognition and charging of tRNA it is important that this base pair be not GC (Yarus *et al.* 1977). Presumably, the added free energy cost of breaking a GC pair makes tRNAs containing a GC at 1–72 less suitable for proper binding to the enzyme.

The 3′5′-exonuclease-active site of *E. coli* DNA polymerase I is observed to denature duplex DNA and bind four single-stranded nucleotides at the 3′ terminus (Steitz *et al.* 1987; Freemont *et al.* 1988). The physiological role of this exonuclease is to 'edit' out mismatched base pairs that are erroneously incorporated at the polymerase active site some 30 Å away. It has been proposed that the specificity of this exonuclease for editing out mismatch base pairs with higher frequency than correctly matched base pairs arises from the former's lower thermal stability (Brutlag & Kornberg, 1972; Steitz *et al.* 1987). In the competition between the duplex-binding polymerase active site and the single-strand binding exonuclease-active site, duplex DNA containing mismatched base pairs will bind to the exonuclease site with greater frequency than correctly matched duplex.

Sequence recognition also arises from the sequence-dependent ability of single-stranded nucleic acid to take up the conformation required for protein binding. The single-stranded acceptor end of $tRNA^{Gln}$ assumes a hairpinned conformation upon binding to GlnRS (Rould *et al.* 1989). This conformation is stabilized by an interaction between the N2 of G_{73} and the backbone phosphate of A_{72}. the inability of the other three bases to form this stabilizing hydrogen bond would be expected to increase the free energy cost of binding $tRNA^{Gln}$ with other bases at this position.

Nucleic-acid distortability as a more indirect source of sequence specificity arises from two facts: (1) Proteins often bind a conformation of a nucleic acid that is altered from its uncomplexed solution conformation. (2) The free energy cost for various nucleic-acid sequences to assume the conformation that is required for its binding to the protein is not the same for different sequences.

2.4 *Role of water molecules in sequence recognition*

Two protein–DNA complexes and one protein–RNA complex are sufficiently well refined at high resolution to show the positions of firmly bound water molecules. Some buried water molecules appear to be playing an important role in both DNA and RNA sequence recognition. Water (or a serine hydroxyl group) can only make a base-specific hydrogen bond if it is also making at least two other hydrogen bonds with obligate donors or acceptors on the protein and is buried form bulk solvent. In this circumstance the two unsatisfied water H-bond donor/acceptors directed towards the nucleic acid become obligate donor/acceptors and become part of the H-bonding template surface of the protein to which the nucleic acid must be complementary for optimal binding. In the trpR–DNA complex there are three water molecules per half operator bound between protein and the DNA bases and two of them appear to be making hydrogen bonds that specify base pairs 5, 6 and 7 from the dyad axis (Otwinowski *et al.* 1988). In this case, water molecules are playing the role of 'honorary' protein side chains. In the 434 repressor complex a water molecule is used or orient the carboxy amide side chain of Gln33 so that the amide is direct towards the O4 of thymine (Aggarwal *et al.* 1988). Water molecules also mediate the interaction of the guanidinium group of Arg43 with AT base pairs in the minor groove.

In the GlnRS complex with tRNA a buried water molecule is an integral part of the H-bonding matrix presented to the shallow groove of the tRNA acceptor stem (Rould *et al.* 1989). Hydrogen bonds between this water molecule and both a buried carboxylate of Asp235 and the backbone amide of residue 183 serve to orient one hydrogen bond donor of the water towards the O2 of cytosine 71 and an acceptor towards the N2 of guanine 2.

2.5 *Role of the minor groove in DNA and RNA recognition*

As pointed out by Seeman *et al.* (1976), thee are fewer features presented by base pairs in the minor groove (as compared with the major groove) that allow discrimination among the two base pairs and their two orientations (Fig. 2). The hydrogen-bond acceptors (N3 on guanine and adenine and O2 on cysteine and thymine) occur in almost identically the same place in the minor groove for all four bases. Only the exocyclic N2 of guanine distinguishes AT from GC and perhaps GC from CG. Furthermore, the minor groove of B-DNA is too narrow to accommodate an α-helix and too deep for the bases to be reached by side chains alone.

It appears, however, that there are ways in which interactions in the minor groove can be sequence specific. The water mediated interactions made by the two symmetry related Arg43 guanidinium groups in 434 repressor stabilize a highly propeller-twisted structure taken up by an AT-rich sequence at the dyad axis of the 434 operator (Aggarwal *et al.* 1988). Since a GC-rich sequence cannot take up this structure it presumably would not be able to make these water-mediated contacts with the protein and would bind less tightly than the AT-rich sequence. In this case sequence specificity is achieved by both water-mediated hydrogen-bonding complementarity and sequence-dependent deformability of DNA.

The sequence preferences exhibited in the DNAse I cleavage of DNA arise from its interactions in the minor groove (Suck *et al.* 1988). The side chain of Tyr that is observed to bind in the minor groove will fit into the normal width minor groove, but not into the narrower minor groove that characterizes AT-rich sequences.

With duplex RNA, which is A-form, the accessibility of the grooves is the reverse of DNA; only the minor or shallow groove is accessible, the major groove being both narrow and very deep. Two sequence-specific interactions between GlnRS and the minor groove of $tRNA^{Gln}$ have been observed (Rould *et al.* 1989). The carboxylate of an aspartic-acid side chain emanating from the amino end of an α-helix interacts simultaneously with the N2 of guanine, the 2′OH of a backbone ribose, and a buried water molecule. One would anticipate that substitution of guanine by any of the other three bases would abolish this

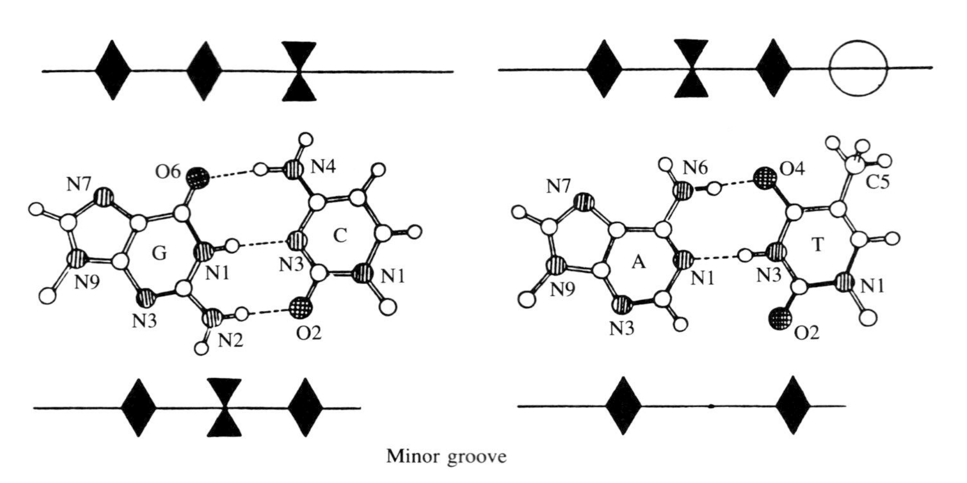

Fig. 2. The hydrogen-bond donors and acceptors presented by Watson–Crick base pairs to the major groove and the minor groove and the minor groove (adapted from Lewis *et al.* 1985). The symbols for hydrogen bond donors (⧗) and acceptors (◆) (Woodbury *et al.* 1989) show a varied pattern presented by the base pairs to the major and an poor information array in the minor groove. While it is possible to distinguish among AT, TA, GC and CG in the major groove, functional groups in the minor groove allow only easy discrimination between AT and GC containing base pairs. ○, methyl group.

interaction thus lowering the affinity, a proposal that is currently being tested. This particular sequence-specific bidentate interaction can only occur with RNA. Another sequence dependent interaction occurs between the peptide backbone carbonyl oxygen of proline and the N2 of quanine, again in the minor groove.

3. DNA-BINDING STRUCTURE MOTIFS

The structures of three different motifs that interact with DNA in a sequence-specific manner are now known: the helix–turn–helix motif characteristic of most prokaryotic regulatory proteins, α-helical structure found in *Eco*R I, and the zinc 'finger' structure found in some eukaryotic transcription factors. In order for a protein to present a surface that is complementary to the major groove of B-DNA, it is necessary for the protein structure to protrude significantly from its own surface. In the case of all three of these motifs it appears likely that an α-helix is utilized in order to present the array of side chains that are complementary to the exposed edges of base-pairs in the major groove of B-DNA. It has been recognized for 30 years that an α-helix will fit into the major groove of B-DNA (Zubay & Doty, 1959) and early models for histone interaction with DNA speculated that an α-helix might fit into the major groove (Sung & Dixon, 1970). In these and later models for *lac* repressor interaction with DNA (Adler *et al.* 1972) and protamine interaction with DNA (Warrent & Kim, 1978) the α-helix was placed parallel to the major groove. It is now recognized that while the α-helix axis may lie parallel to the groove as model building suggested to be the case in the *cro*–DNA complexes (Ohlendorf *et al.* 1982), it may in fact penetrate the major groove in any one of several different orientations. The α-helix axis is almost perpendicular to the DNA helix axis in the case of *Eco*R I (McClarin *et al.* 1986) and *trp* repressor (Otwinowski *et al.* 1988) or parallel to the DNA base pairs as in CAP (Weber & Steitz, 1984), λ repressor (Jordan & Pabo, 1988) or 434 repressor (Aggarwal *et al.* 1988). The principle general feature emerging from all of the known complexes, however, is that most of the direct sequence-specific interactions with bases are made by amino-acid side chains emanating from the amino end of α-helices that penetrate the major groove of B-DNA. This also places the positive end of the helix electrostatic dipole in the major groove.

3.1 *Helix–turn–helix*

The helix–turn–helix motif that is now found to be important in DNA sequence recognition in many repressors and activators was initially identified as a common recognition element from a comparison of the structures of CAP and *cro* (Steitz *et al.* 1982). Amino-acid sequence similarities between these two proteins and among a wide variety of other repressors suggested that this helix–turn–helix structure might occur rather generally (Anderson *et al.* 1982; Matthews *et al.* 1982; Sauer *et al.* 1982; Weber *et al.* 1982*a*). Subsequent determination of the structures of λ repressor, *trp* repressor and 434 repressor and comparison of their structures with those of CAP and *cro* have shown that this motif structure is indeed highly conserved (Ohlendorf *et al.* 1983; Zhang *et al.* 1987). Comparison of pairs of helix–turn–helix structures show that the α-carbon atoms superimpose on each other within an RMS distance ranging from 0·7 to 1·0 Å (Zhang *et al.* 1987). Sequence similarities suggest that the helix–turn–helix motif is found in some eukaryotic regulatory proteins as well, most notably in the homeo domain of the mouse and *Drosophila* homeotic proteins (Laughon & Scott, 1984; Shepherd *et al.* 1984) which has now been confirmed by a high-resolution 2D NMR structure determination of the 60-residue *Antennapedia* homeodomain from *Drosophila* (Qian *et al.* 1989).

What accounts for the structural conservation of this two-helix motif? There are 6 amino acids out of 21 whose nature is conserved to varying degrees among the related sequences (Fig. 3). Four of these residues make important hydrophobic contacts between the two helices assuring the preservation of their mutual orientation; two other residues, an alanine and a glycine, are important for the bend between the two helices (Weber *et al.* 1982*a*; Anderson *et al.* 1982). Most of the rest of the residues are hyper-variable among the various proteins because they are either facing the differing DNA sequences or the remaining portion of the repressor proteins that differs from protein to protein. It is not clear why the structure of the helix–turn–helix motif is so highly conserved in spite of the fact

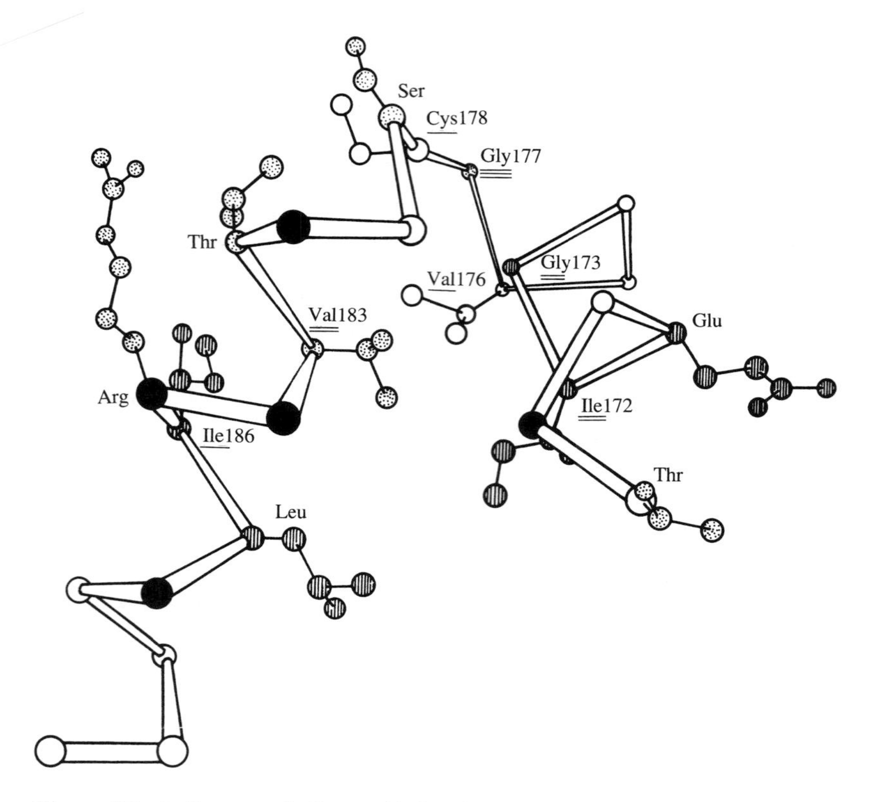

Fig. 3. The helix–turn–helix motif showing the positions of conserved hydrophobic residues that form the contacts between the two helices and the conserved residues at the bend between the helices (reproduced from Weber *et al.* 1982*a*). The sequence and numbers are those of CAP. The most highly conserved residue (Gly177) is triply underlined, usually conserved doubly underlined and less conserved singly underlined. This structure is stabilized by the hydrophobic interactions between the side chains of the conserved Val183, Ile172, Val176 and Ile186. Stippled side chains are identical in CAP, *gal* and *lac* repressors while striped side chains are closely similar.

that the relative orientation of this motif and the DNA is not (Fig. 4). Possibly this motif allows an energetically stable way of protruding an α-helix from the surface of the protein so that it might interact in the major groove of DNA. It is also possible that this motif allows for the very wide variation in the amino-acid sequences required to recognize different DNA sequences without damaging the motif structure.

3.2 *Zinc 'fingers'*

A second DNA-binding motif first identified as a 30-residue repeating-sequence element in transcription factor IIIA from *Xenopus* and now found in a very large number of eukaryotic transcription factors was called a zinc 'finger' because of a

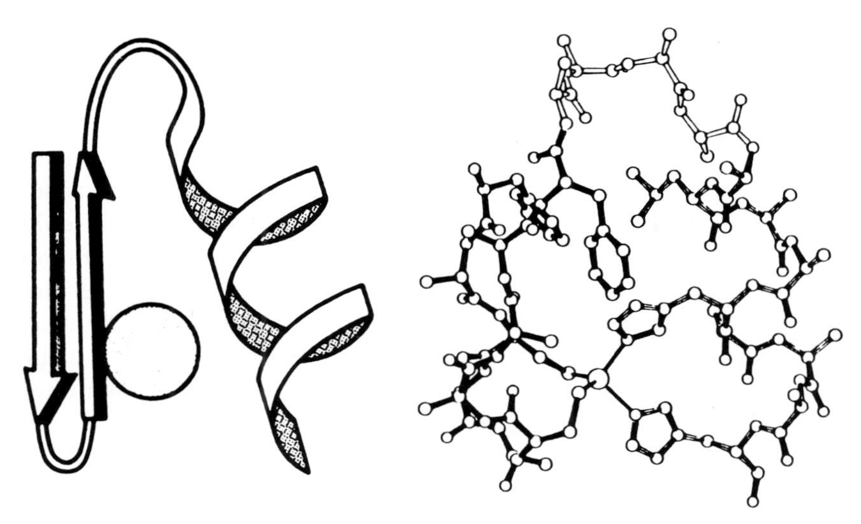

Fig. 5. Schematic drawing (left) and more detailed structure including conserved residues (right) of the model of a zinc 'finger' proposed by Berg (1988) consisting of two anti-parallel β-strands and an α-helix held together by a buried zinc ion liganded to two Cys residues on the β-strands and two His side chains on the α-helix (reproduced from Berg, 1988). 2D NMR data now confirm the essential features of this model (Parraga *et al.* 1988; Lee *et al.* 1989).

firmly bound zinc ion essential to its structure (Miller *et al.* 1985). Both model building, employing a clever use of the structural data bases (Berg, 1988), and 2D NMR measurements (Parraga *et al.* 1988; Lee *et al.* 1989) have separately suggested an approximate structure for these zinc fingers consisting of a two-strand anti-parallel β-sheet and an α-helix surrounding a centrally bound zinc ion (Fig. 5). Evidence on how this motif interacts with DNA is still missing though one hypothesis proposes that some of the sequence-specific recognition is achieved through side chains emanating from the α-helix that is interacting in the major groove of the DNA (Berg, 1988). Since TFIIIA interacts with 5S RNA and the same sequences in duplex DNA, one might expect it to interact with both RNA and DNA in the same fashion. DNA can assume A-form more easily than RNA can assume B-form, suggesting that TFIIIA interacts with A-form DNA and RNA. Since only the minor groove of standard A-form is accessible to an α-helix, it is possible that TFIIIA binds in the minor groove of both DNA and RNA. Alternatively, the A-form structures with which this protein interacts may be sufficiently distorted to allow interaction in the major groove, as suggested by the crystal structure of the TFIIIA DNA-binding site (McCall *et al.* 1986). A related DNA-binding domain that contains Zn has four Zn-liganding cysteines rather than two His and two Cys. Its structure is probably different from that of the TFIIIA family.

3.3 *Helices of the dinucleotide fold*

Two enzymes of known structure appear to achieve sequence recognition through the use of α-helices that are part of a parallel β-sheet motif that is often called a dinucleotide (or Rossmann) fold. The amino ends of these recognition α-helices interact in the major groove of B-DNA in the case of *Eco*R I (McClarin *et al.* 1986) and the shallow groove of RNA in the case of glutaminyl-tRNA synthetase (Rould *et al.* 1989). The dinucleotide fold is a common parallel β-sheet structure found in all of the dehydrogenases to bind the NAD and also observed to bind ATP and GTP in kinases and G proteins (Rossmann *et al.* 1975). In *Eco*R I two of the α-helices associated with this motif are penetrating the major groove of DNA while in GlnRS one α-helix from this motif is seen to provide interaction in the minor or shallow groove of tRNA. The axes of the recognition α-helices of *Eco*R I are almost perpendicular to the DNA-helix axis. Unlike the helix–turn–helix and Zn-finger motifs, there is no detailed structural or sequence similarity between these two proteins.

3.4 *Other motifs*

Experiments from the laboratories of Pabo and Sauer have shown that the extreme amino terminus of λ repressor is important for this protein's ability to discriminate among DNA sequences (Pabo *et al.* 1982). From high-resolution crystal structure of the complex it appears that an extended amino terminal arm is making sequence-specific interactions with bases via the major groove (Jordan & Pabo, 1988; Sauer *et al.*, 1990; Pabo, private communication, 1989). Apparently, potentially flexible chain termini can penetrate the major groove and make discriminating contacts.

It was proposed on the basis of general model-building by Carter & Kraut (1974) and Church *et al.* (1977) that anti-parallel β-strands could interact in the grooves of DNA and RNA (Fig. 6). There are now two proteins that appear to use anti-parallel β-structure in sequence discrimination. *E. coli* integration host factor (IHF) is a small sequence-specific DNA-binding protein essential for numerous processes including the integrating of λDNA into host *E. coli* DNA (reviewed in Friedman, 1988). The two similar subunits of this dimeric protein each show sequence homology to *E. coli* Hu. In order to account for DNA protection data, it has been proposed that IHF makes its sequence-specific interactions in the minor groove of B-DNA using side chains emanating from two arms of two-stranded anti-parallel β-structure that encircle the DNA (Yang & Nash, 1989). No detailed model of how this is accomplished yet exists.

The recently determined crystal structure of *E. coli met* repressor (Rafferty *et al.* 1989) shows none of the previously observed motifs, but rather (according to a note added in proof) a co-crystal structure shows anti-parallel β-strands that fit into the major-groove of B-DNA (Rafferty *et al.* 1989).

It is clear from the sequences of many eukaryotic transcription factors and enhancers that other DNA-binding motifs must exist. Another family of related

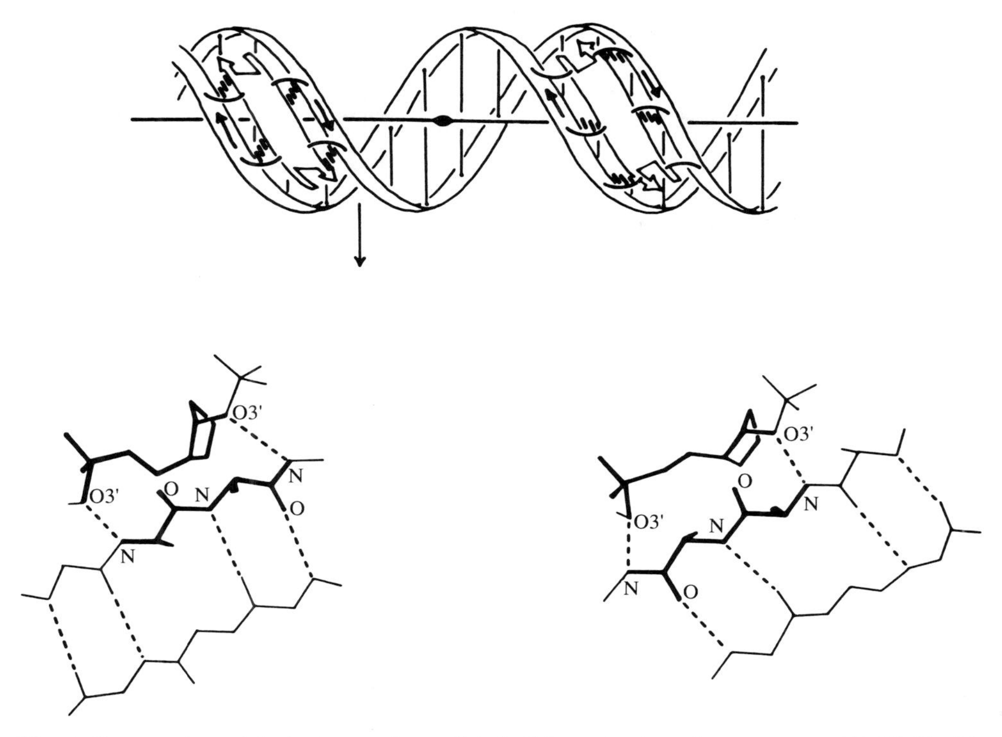

Fig. 6. Proposed model placing anti-parallel β-ribbons in the minor groove of B-DNA. The top and bottom present the two different orientations of the N–C direction of the polypeptide and the 3′–5′ direction of the DNA (from Church *et al.* 1977).

proteins has been identified from sequences that are characterized by 4 or 5 leucines that repeat every 7 amino acids (Landschultz *et al.* 1988). However, nothing is currently known about the structure of the region of these proteins that interacts with DNA. It will be interesting to see whether each of the various motifs of this and other sequence-specific proteins employ different strategies of presenting an α-helix into the major groove of B-DNA or whether other secondary structures such as β-turns or anti-parallel β-strands are utilized in direct interaction with bases.

4. SIMILARITIES AND DIFFERENCES IN RNA AND DNA RECOGNITION

Direct recognition of base sequence can be achieved in both duplex RNA and DNA through penetration of grooves by α-helices. However, while most direct sequence recognition of DNA is achieved via interactions in the major groove of B-form DNA, the major groove of double stranded RNA (which is A-form) is too narrow and deep to allow penetration by an α-helix without extensive distortion. In contrast, the minor or shallow groove of RNA accessible to α-helices anti-parallel β-strands and offers some recognition opportunities available only to RNA, e.g. bidentate interaction with both the N2 of guanine and the 2′OH of ribose.

Recognition of duplex RNA via the minor groove may or may not permit discrimination among all four base pairs. Both AU and UA present identically the same hydrogen-bond acceptors in the minor groove, though the bifurcated hydrogen bonding possible for oligo-adenosine tracts (Nelson *et al.* 1987) may be adequately different from mixed (AU) sequences. Likewise, the N2 of guanine is situated very close to the pseudo-dyad axis of a GC base pair so that it may not be possible to distinguish GC form CG via the minor groove. Further, GU base pair which occur in RNA but not DNA may well be distinguishable from both AU and GC base pairs.

Since RNA molecules can take on more varied overall 3D shapes than duplex DNA, discrimination among a range of shapes presents an easy recognition problem. Within a family of RNA molecules of identical structure, such as tRNA, single-stranded regions present recognition opportunities not present in DNA. Bases in the anti-codon loop of $tRNA^{Gln}$ are directly recognized by GlnRS, for example (Rould *et al.* 1989).

In both RNA and DNA recognition the free energy 'cost' of a particular nucleotide sequence to assume the structures required for binding to the protein plays a significant role. Both duplex DNA and RNA as well as single-stranded RNA show a sequence dependence to their 'deformability' from their uncomplexed structure. As a final point of similarity between RNA and DNA recognition, buried water molecules appear to function as surrogate protein side chains and form part of the interaction surface to which the nucleic-acid sequence must be complementary.

5. SEQUENCE-SPECIFIC DNA-BINDING PROTEINS

5.1 *Repressors and activators*

5.1.1 Lac *operon regulation*

Transcription of the three genes of the lactose operon is controlled by both a negative regulator (the *lac* repressor protein) and a positive regulator (the catabolite gene activator protein, CAP) (Miller & Reznikoff, 1978). Only in the presence of lactose and in the absence of glucose does transcription proceed at a high level. In that event allolactose, the molecule that functions as an inducer, increases in concentration and binds to the *lac* repressor resulting in its dissociation from DNA; cAMP also increases in concentration and binds to CAP which then binds to the DNA and stimulates RNA polymerase.

(*a*) E. coli *catabolite gene activator protein* (*CAP*)

The catabolite gene activator protein (CAP) from *E. coli*, also known as the cyclic AMP receptor protein (CRP), functions primarily as an activator of transcription from numerous operons (de Crombrugghe *et al.* 1984) although it also can function as a repressor (Aiba, 1983). When glucose levels drop, cyclic AMP (cAMP) levels rise in *E. coli* and cAMP binds to CAP. cAMP–CAP complex binds in a sequence-specific manner to the upstream region of numerous operons and

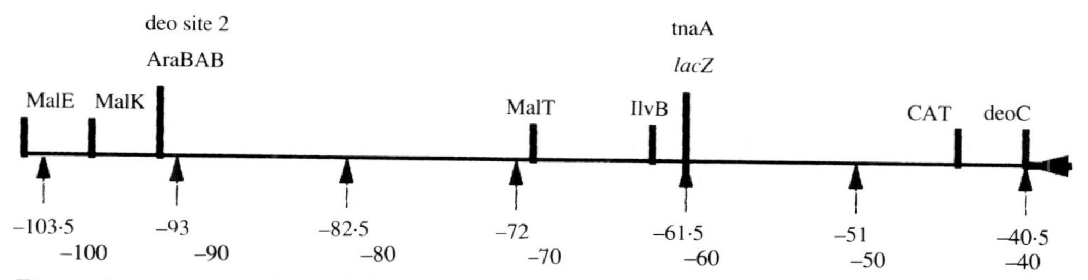

Fig. 7. Base-pair positions of the centres of various CAP-binding sites in operons that it regulates relative to the transcription start site. Arrows mark 10·5 bp from the centre of the CAP-binding site in the *lac* operon. Note that most CAP sites are on the same side of the DNA as the polymerase.

activates transcription by RNA polymerase. The three questions being addressed by studies of this gene regulatory protein are: (1) what structural change is induced in CAP by the binding of cAMP that results in the sequence-specific DNA binding? (2) How does CAP recognize the specific DNA sequence, a question raised with all of the sequence-specific biding proteins? (3) How does the binding of CAP to specific DNA sequences activate RNA polymerase transcription? The latter question is made more challenging by the observation that the binding sites for CAP in the different promoters that it activates occur at several different distances to the transcription start site in each case (Steitz & Weber, 1985; de Crombrugghe *et al.* 1984). The centre of the CAP-binding site of DNA is located variously between approximately −41 and −103 in the operons that it activates (Fig. 7). Since it is commonly assumed, though not established, that CAP activates the polymerase by direct protein–protein interaction, the puzzle that is presented by the CAP system is to discover a common mechanism of activation that is consistent with the various observed CAP-binding sites. This variable distance along the DNA between the CAP-binding sites and the transcription start is reminiscent of the variable relationships that exist between eukaryotic transcription activators and transcription start sites.

CAP is a dimer of 22500 molecular-weight-identical subunits. The crystal structure determined by McKay & Steitz (1981) showed that the protein is folded into two domains (Fig. 8). The larger *N*-terminal domain forms a β-roll structure into which a molecule of cAMP is observed to bind. The smaller C-terminal domain contains three α-helices, two of which were found to be identical to helix two and three in *cro* (Steitz *et al.* 1982) and subsequently to the other bacterial regulatory proteins. The initial structure determination by McKay and Steitz was done in the absence of the amino-acid sequence which was subsequently determined by Aiba *et al.* (1982) and Cossart & Gicquel-Sanzey (1982). The amino-acid sequence was incorporated into the structure (McKay *et al.* 1982*b*) that has been refined at 2·5 Å resolution to a crystallographic *R*-factor of 0·21 (Weber & Steitz, 1987).

The CAP–cAMP dimer (Fig. 9) is asymmetric in the crystal of the cAMP complex (though symmetric in the DNA co-crystal) and is held together entirely by interactions between the two cAMP-binding domains (McKay & Steitz, 1981;

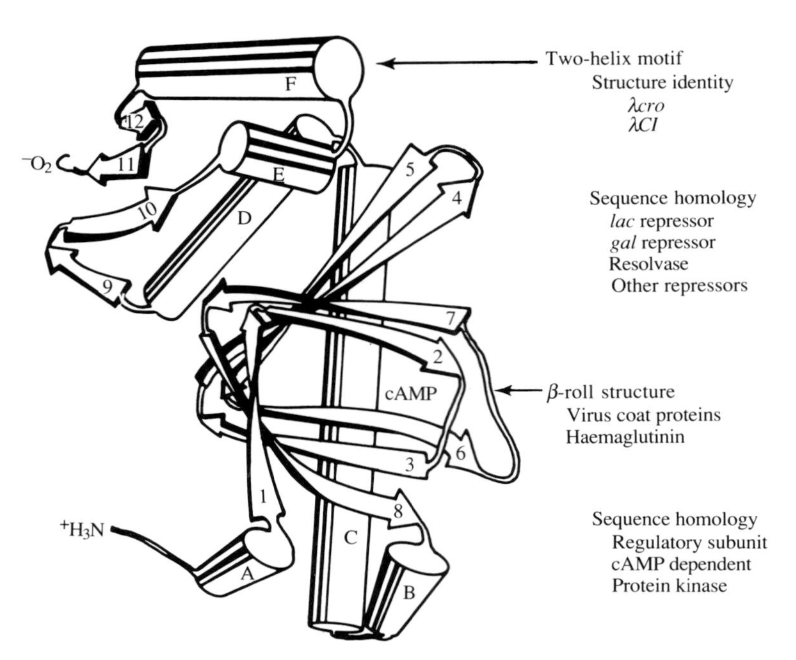

Fig. 8. A schematic drawing of the CAP monomer showing the two domains. The larger *N*-terminal domain is seen to bind cAMP while the C-terminal domain contains the helix–turn–helix common to bacterial regulatory proteins (from McKay *et al.* 1982*b*).

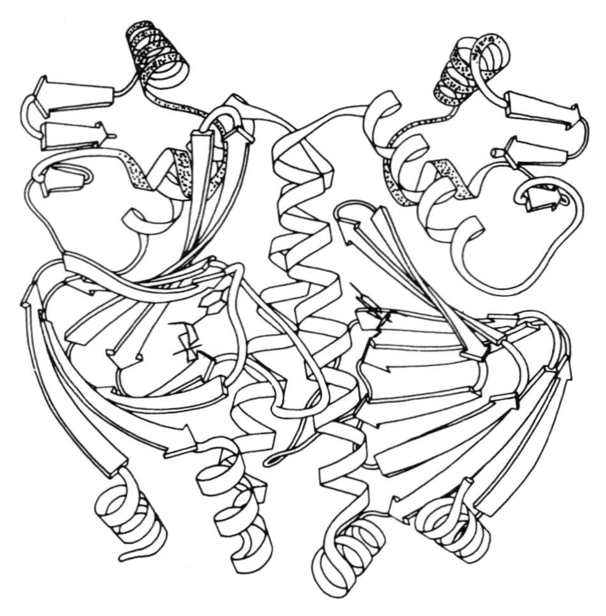

Fig. 9. A drawing of the CAP protein dimer showing the location of the two bound cAMP molecules and the positions of the two helix–turn–helix motifs (dotted). The drawing is a modification of one made originally by Jane Richardson and published by Steitz *et al.* (1982).

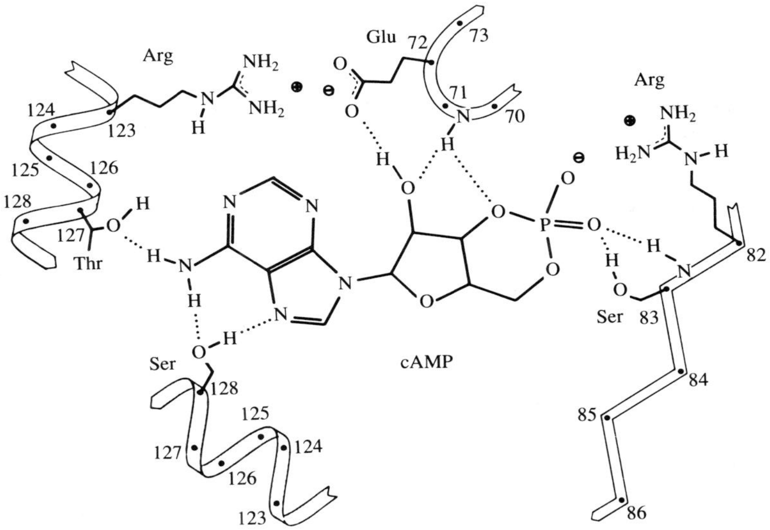

Fig. 10. Schematic drawing of the H-bonding interactions between CAP and cAMP (from Weber & Steitz, 1987).

McKay *et al.* 1982*b*). The relative orientation of the two domains differs in the two subunits being more 'open' in one subunit than the other. Thus, the dyad axis that relates the two DNA-binding domains is tilted by 18° from the dyad axis relating the cAMP-binding domain (McKay & Steitz, 1981). The majority of contacts holding the two subunits together are provided by the two C α-helices (Figs 8 and 9) of the cAMP-binding domain. These α-helices interact in a coiled-coil fashion as is now proposed for the eukaryotic transcription factors GCN4, *fos*, *Jun* and others that contain a periodic Leu repeat (O'Shea *et al.* 1989).

(i) *cAMP binding*. In the crystal one molecule of cAMP is observed in each subunit of the dimer bound in the *anti* conformation. Attempts to fit the *syn* conformation to either the initial experimental electron density map or an electron density map calculated from the refined coordinates in which the cAMP has been omitted were not successful. However, Gronenborn & Clore (1982) concluded from transferred NOE measurements that cAMP binds to CAP in a *syn* conformation. Although the source of this discrepancy between the conclusions drawn from X-ray and NMR measurements has not been established, it is possible that the high concentrations of cAMP (3·3 mM) in D_2O required for the NMR measurements led to the binding of cAMP to another weaker site.

The cAMP is nearly completely buried in the interior of CAP and all of its hydrogen bond donors and acceptors are making interactions with the protein (Fig. 10). Of particular importance, both from analogue-binding studies (Scholübbers *et al.* 1983; Ebright *et al.* 1985) and in the crystal structure, is the six amino group of adenine seen to be interacting both with Ser125 and Thr127 from the other subunit. The observation that cAMP is binding at the subunit interface and interacting with both may be of significance for its allosteric activation of DNA binding.

Mutant CAP proteins have been isolated that activate transcription in *E. coli* strains that lack adenyl cyclase (Aiba *et al.* 1985; Garges & Adhya, 1985). These mutant proteins, termed CAP*, can bind to specific DNA sequences in the absence of cAMP and activate transcription. Presumably, these point mutations affect the equilibrium between the conformation of CAP that is able to bind DNA in a sequence specific manner and the conformation that is not. In the CAP structure these mutations are located either at the interface between the large and small domains (mostly on α-helix D) or at the interface between the two subunits of the dimer. Crystallographic studies of a few CAP* mutants crystallized in the cAMP crystal form show very little change in the structure of the cAMP complex (Shields, Engelman & Steitz, unpublished; Weber, personal communication).

The conformational change induced in CAP by the binding of cAMP has not been directly established since crystals of the cAMP free CAP suitable for crystallographic studies have only recently been obtained (Schultz & Steitz, unpublished). McKay *et al.* (1982*b*) proposed that the binding of cAMP alters the relative orientations of the DNA- and cAMP-binding domains and/or the relative orientation of the two subunits of the dimer. Such conformational changes would create and destroy the sequence-specific DNA-binding site that is believed to span two subunits (see below). This proposal appears to be consistent with the location of the CAP* mutations. It is also analogous to the conformational change observed to be produced by the binding of tryptophan to the *trp* repressor (Zhang *et al.* 1987) (see below).

The cAMP-binding domain of CAP shows an extensive amino-acid sequence similarity to the regulatory subunit of the cAMP- and cGMP-dependent protein kinases (Weber *et al.* 1982*b*). A detailed model of the cAMP-binding domains of the RII subunit of the cAMP-dependent protein kinase has been built (Weber *et al.* 1987) that appears to correlate well with most of the known biochemical and functional data on this protein.

(ii) *Model building of CAP–DNA complex.* Model building of DNA on CAP was greatly aided by electrostatic calculations of the positive and negative charge potential on the CAP protein (Steitz *et al.* 1982; Steitz *et al.* 1983; Weber & Steitz, 1984; Warwicker *et al.* 1987). As part of the initial structure determination (McKay & Steitz, 1981) it was proposed that CAP might bind to a left-handed B-type DNA with the F-helices of the two subunits lying parallel to the major grooves in a manner analogous to a model proposed at the same time for cro binding to right-handed DNA (Anderson *et al.* 1981). This model of CAP binding to left-handed DNA was demonstrated to be incorrect when Kolb & Buc (1982) found that CAP binding to closed circular DNA did not induce the large unwinding in the DNA predicted by the model. Subsequent incorporation of the amino-acid sequence into the CAP structure allowed calculation of the positive electrostatic charge potential that was used to orient DNA on CAP (Steitz *et al.* 1982). A detailed model, constructed by Weber & Steitz (1984) (Steitz *et al.* 1983), placed the amino ends of the protruding F α-helices into the major groove of right-handed B-DNA but more nearly parallel to the bases rather than parallel to the groove, as with the *cro*–DNA model.

Some of the specific interactions proposed between protein side chains on the F α-helix and the exposed edges of base pairs in the major groove are consistent with subsequent genetic studies (and the preliminary co-crystal structure) and some are not. Nevertheless, taken together, the genetic data (Ebright *et al.* 1984; Gent *et al.* 1987; Ebright *et al.* 1987) support the general orientation of DNA on CAP as suggested by the electrostatic calculations as does the preliminary co-crystal structure of a CAP–DNA complex (Schultz *et al.* 1990). Comparison of the model built complexes of λ repressor and *trp* repressor with the crystal structures of these proteins bound to their operators shows that the overall orientation of the DNA on the protein as well as some specific interactions were correctly guessed (Jordan & Pabo, 1988; Otwinowski *et al.* 1988). However, it was apparently not possible to correctly anticipate all of the interactions or the detailed changes in protein and DNA structure that occur upon complexation. Doubtless, the same will hold true for the model built CAP–DNA complex.

(iii) *DNA bending by CAP and its role in sequence specificity.* A variety of experimental measurements have been interpreted to indicate that CAP introduces a significant bend into DNA when it binds in a sequence-specific fashion. CAP complexed with a restriction fragment containing a CAP-binding site shows an anomalous electrophoretic mobility on polyacrylamide gels that is dependent on the distance of the CAP site from a DNA end; this mobility change has been interpreted as resulting from CAP-induced DNA bending (Wu & Crothers, 1984). Electric dichroism studies of CAP bound to DNA fragments in solution have also suggested that CAP introduces a significant bend in the DNA (Porschke *et al.* 1984). Electron micrographs of CAP bound to DNA fragments show bends in the DNA at the point that CAP contacts the DNA, although these micrographs were interpreted to suggest that CAP bends the DNA away from the protein (Gronenborn *et al.* 1984). A more likely interpretation is that the DNA is bent around the protein to a significant extent.

The length of the DNA site to which CAP binds appears to be at least 28–30 bp, larger than the 18–20 bp typically observed with other bacterial repressors. Ethylation interference studies show that ethylation of DNA backbone phosphates spanning 24 bp affect CAP binding (Majors, 1977). Studies of the relative affinity of CAP for DNA fragments of various lengths establish that 28–30 bp are required to achieve the nearly full affinity of CAP for DNA (Liu-Johnson *et al.* 1986).

The two observations of CAP-induced bending of DNA and the unusually large DNA-binding site appear to be correlated in an interesting way. A plausible way to extend the size of the DNA with which CAP can interact in the relative orientation suggested by genetic and electrostatic studies is to introduce a significant bend into the DNA. Weber & Steitz (1984) had smoothly bent the DNA to a radius of curvature of 70 Å to extend the binding site to about 20 bp. However, in order to account for a DNA-binding site size of 28 bp and a positive electrostatic potential field that forms a ramp on three sides of CAP a more sharply bent DNA is required (Warwicker *et al.* 1987) (Fig. 11). Kinking the DNA at 6 positions maintains its favourable electrostatic contact with the protein over 30 bp and results in an overall DNA bend of about 140° (Fig. 12). Contact between the DNA and protein includes interactions with the cAMP-binding domain.

Gartenberg & Crothers (1988) have shown an experimental correlation between the nucleotide sequences at putative bend loci and both the affinity of CAP for DNA and the extent of DNA bending. They found using gel mobility assays that substitution of GC-rich for AT-rich sequences at nucleotides 10 and 11 from the dyad axis (where the minor groove faces the protein) results in reduced bending and reduced affinity. In contrast, at nucleotide 16 from the dyad axis GC-rich sequences favoured bending and binding. These observations led them to conclude that GC-rich sequences favour major groove compression (kinking into the major groove) whereas AT-rich sequences favour minor groove compression, a conclusion drawn earlier by Drew & Travers (1984) and Satchwell *et al.* (1986) from their statistical analyses of DNA sequences bound to nucleosomes. Warwicker *et al.* (1987) showed that the sharp DNA bending required to maintain its contact with the protein resulted in greatly increased calculated electrostatic stabilization of the complex. Thus, only CAP-binding sites that contain sequences that facilitate DNA binding will show the strong favourable electrostatic interaction in the protein. This is another example in which sequence specificity arises not through direct interaction of the protein with the specific sequence, but rather through the ability of certain sequences to adapt the conformation required by the protein for optimal DNA binding.

(iv) *Structure of CAP co-crystallized with DNA shows large DNA kinks.* Large, crystallographically suitable crystals of CAP complexed with cAMP and 30 bp of duplex DNA containing a single base 5′ overhang have been grown and diffract isotropically to better than 3 Å resolution (Schultz *et al.* 1990). Early attempts to co-crystallize CAP with either a 16 or a 22 bp fragment of DNA yielded only crystals of cAMP with CAP in spite of the fact that both DNA and protein were present at concentrations approaching 0·1 mM and considerably in excess of even the non-specific dissociation constant (Goldman & Steitz, unpublished, 1983; Steitz, unpublished, 1985). However, data from the Crothers' laboratory (Liu-Johnson *et al.* 1986) showed that 28–30 bp were required for full binding affinity. Thus, crystallization was explored using specific sequence DNA oligonucleotides of lengths varying from 28 to more than 40 bp (Schultz *et al.* 1990). While many different complexes formed crystals, the best were obtained using a 30 bp symmetric DNA with 5′G overhanging nucleotides.

Structure determination and preliminary refinement (Schultz, Shields, Steitz, unpublished, 1989) of these CAP–DNA co-crystals at 3 Å resolution provide results that are consistent with the overall features of the CAP–DNA model of Warwicker *et al.* (1987). The co-crystal structure was solved by placing the coordinates of the CAP–cAMP structure previously solved into the unit cell of the DNA complex using standard rotation and translation function procedures. However, it appears that the dimer is symmetric in this complex and consists of two 'closed' subunits. Using difference Patterson methods, the positions of four bromine atoms were located in a complex with DNA containing four 5BrdU at the ends of DNA. The positions of these Br atoms are within a few Angstroms of the positions expected from the CAP–DNA model of Warwicker *et al.* (1987) symmetrized with 'closed' subunit. Furthermore, an electron density map calculated using phases from the protein alone shows DNA bent around the protein as predicted by model building.

The present partially refined structure while similar overall to the earlier model-built complexes (Steitz *et al.* 1983; Warwicker *et al.* 1987) shows some significant differences. The protruding F α-helices fit more deeply into the major groove and the local distortions in the DNA structure are larger and more varied than anticipated. Further, there is extensive contact between the protein and the DNA backbone at the dyad axis. It appears that DNA bound to CAP is indeed highly bent with an overall bend angle of about 90°. The overall bend is achieved largely through two 45° kinks produced by base-pair rolls between base pairs 5 and 6 from the dyad axis (GC and TA, respectively). This results in a closing of the major groove and an enormous widening of the minor groove. The DNA is bent away from the protein at its ends in order to achieve end to end stacking of DNA in the crystal. It is likely that longer DNA molecules will be even more highly bent than this 30-mer due to additional electrostatic interactions that can then be made with the cAMP-binding domain. Both model building (Warwicker *et al.* 1987) and the data of Gartenberg & Crothers (1988) imply that additional kinks should occur between base pairs 10 and 11 and also between base pairs 15 and 16.

(v) *Mechanism of CAP activation of transcription.* While it is not clear what role the CAP-induced DNA bending plays in its activation of RNA polymerase, it certainly changes the relative orientations of CAP and the polymerase when compared with straight DNA (Warwicker *et al.* 1987). It seems likely that in the *lac* operon the cAMP-binding domain is adjacent to the polymerase rather than the DNA domain (Fig. 13). It is not possible that polymerase interacts directly with the DNA-binding domain as proposed by Irwin *et al.* (1987) because (1) Glu191 (whose mutation affects activation) is buried between protein domains in this CAP–DNA complex and (2) Glu191 lies a significant distance from the polymerase-binding site.

It remains possible that CAP-induced bending of DNA brings upstream DNA into contact with RNA polymerase by producing a DNA loop. Formation of such a loop could occur by CAP binding at the −41, −61, −71 and −103 distances from the transcription start at which it is known to function (Fig. 13). If a CAP-induced interaction of the upstream DNA with polymerase is at least partially responsible for transcription activation then a similar interaction could be induced by CAP binding to each of these sites.

(*b*) E. coli lac *repressor protein*

The *lac* repressor protein, in addition to CAP, is required for the regulation of transcription from the *lac* operon in *E. coli* (for reviews see chapters in Miller & Reznikoff, 1978). In the absence of galactose, the *lac* repressor binds to the *lac* operator centred at +10·5 from the transcription start and blocks RNA polymerase. In the presence of galactose but in the absence of glucose transcription can proceed from the *lac* operon. In this case, allolactose binds to the

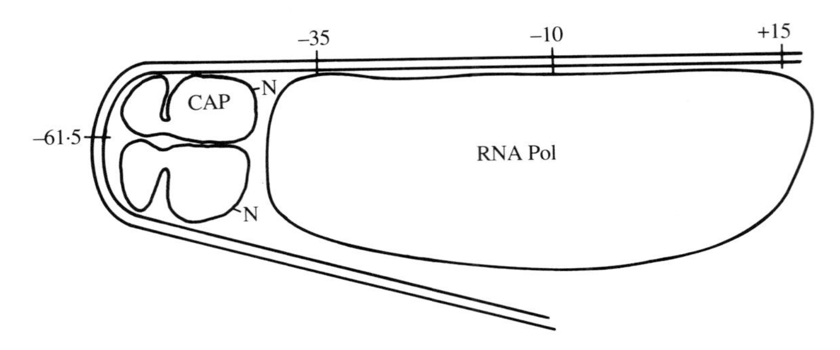

Fig. 13. Schematic drawing showing the relative orientation between CAP on bent DNA and RNA polymerase (from Warwicker *et al.* 1987). It seems plausible that the cAMP-binding domain may contact the polymerase. It also seems plausible that DNA upstream of the CAP site might contact the plymerase in an interaction that is important for polymerase activation. With such a model for activation CAP sites further from the polymerase could make larger loops that still have upstream DNA interacting with the polymerase.

lac repressor causing it to be released from the *lac* operator thus relieving its inhibition of transcription by RNA polymerase.

Until the advent of crystallographic studies of gene regulatory proteins, the *lac* repressor was perhaps the best understood of these gene regulatory molecules (Müller-Hill, 1975; Dickson *et al.* 1975). It is a tetramer of identical 38000 molecular-weight subunits. The sequence-specific DNA-binding activity is confined to a small amino terminal domain while the allosteric activator and subunit interaction is a property of the larger C-terminal domain (Miller, 1978; Ogata & Gilbert, 1978). This two-domain, two-activity feature first demonstrated for *lac* repressor (Adler *et al.* 1972; Platt *et al.* 1973; Files & Weber, 1976) is now commonly observed with many of the allosterically controlled prokaryotic gene regulatory proteins, such as λ*cI* repressor, *lex*A, catabolite gene activator protein, resolvase and Gal repressor, to name but a few, and genetic studies of eukaryotic transcription factors establish similar architectural principles. In this case of the *lac* repressor mutations affecting specific DNA binding are confined to the amino terminal 60 residues (Adler *et al.* 1972; Miller, 1978). Mild proteolytic cleavage of the protein using trypsin results in 51 or 59 residue amino terminal 'headpiece' that still binds to DNA in a sequence-specific fashion and a 30000 molecular-weight carboxy terminal 'core' protein that retains the ability to bind the inducer, isopropylthiogalactoside (IPTG), and to form a tetramer (Platt *et al.* 1973; Files & Weber, 1976; Geisler & Weber, 1977). Although both the *lac* repressor and the 'core' proteins were crystallized in the early and middle 1970s (Steitz *et al.* 1974; 1979), the structure of this large tetrameric molecule has not yet been established by X-ray crystallographic methods.

Nevertheless, the overall shape, the arrangement of subunits in the location of the DNA-binding domains has been derived from a combination of small angle solution X-ray (Mckay *et al.* 1982*a*) and neutron (Charlier *et al.* 1980) scattering and analysis of *lac* repressor crystals. Electron microscopic analysis of crystals of *lac* repressor suggested that the molecule is 120–140 Å long (Steitz *et al.* 1974).

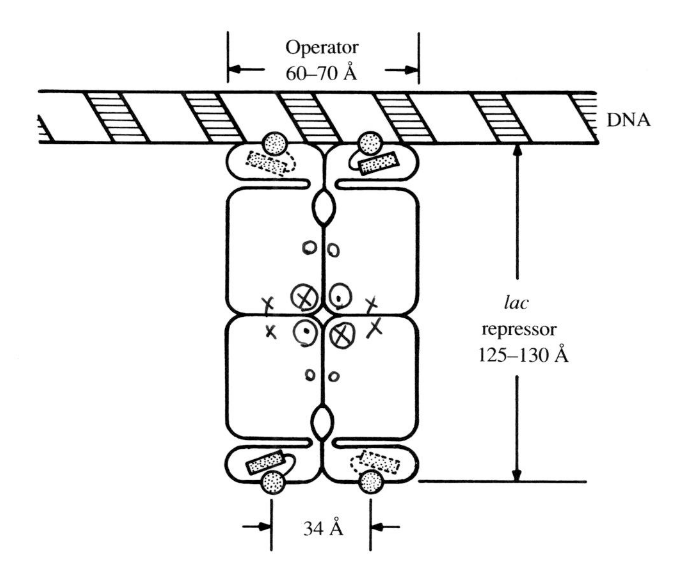

Fig. 14. Schematic drawing of the *lac* repressor tetramer structure as derived from small-angle X-ray and neutron-scattering experiments in solution (McKay *et al.* 1982*a*; Charlier *et al.* 1980). The stippled areas indicate the anticipated positions of the helix–turn–helix motifs (from Weber *et al.* 1982*a*).

Since one crystal form of the *lac* repressor core has a cell dimension of 35 Å (McKay & Steitz, unpublished), it appears likely that one molecular dimension is approximately 35 Å. Consistent with such an asymmetric shape is the radius of gyration measured for the intact repressor. The location of the DNA-binding domains relative to the 'core' was determined by measuring the radius of gyration of the core as well as that of the intact repressor and applying the parallel axis theorem (Charlier *et al.* 1980; McKay *et al.* 1982*a*). The centre of scattering mass of the DNA-binding domains was calculated to be 60 Å from the centre of scattering mass of the intact molecule. To accommodate crystal packing requires placing the DNA-binding domains in pairs at the long end of a rectangular planar arrangement of subunits (Fig. 14). A rotation function calculated using low-resolution X-ray diffraction data from crystals of the *lac* repressor core suggests 222 symmetry for the tetramer (Steitz *et al.* 1979). Although in principle the subunits could be arranged in a tetrahedral fashion, to accommodate a unit cell dimension of 35 Å requires a rectangular planar arrangement. This low-resolution model of the *lac* repressor tetramer looks strikingly like two CAP dimers held together at the end opposite the DNA binding domains. Recently, deletion of C-terminal residues has resulted in a dimeric repressor that binds equivalently to operator DNA (Müller-Hill, private communication).

The amino terminal DNA-binding domain of *lac* repressor was shown by Matthews *et al.* (1982) to exhibit sequence similarities to the helix–turn–helix of other repressor and activator proteins, while the C-terminal core domain was found by Müller-Hill (1983) to exhibit a strong sequence similarity to the galactose-binding protein as well as other periplasmic sugar-binding proteins in

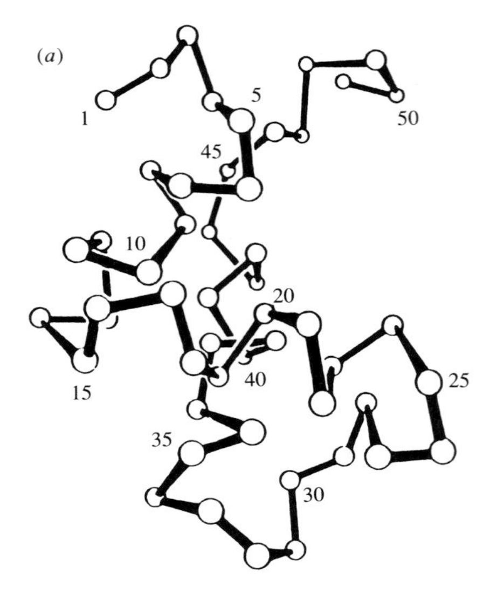

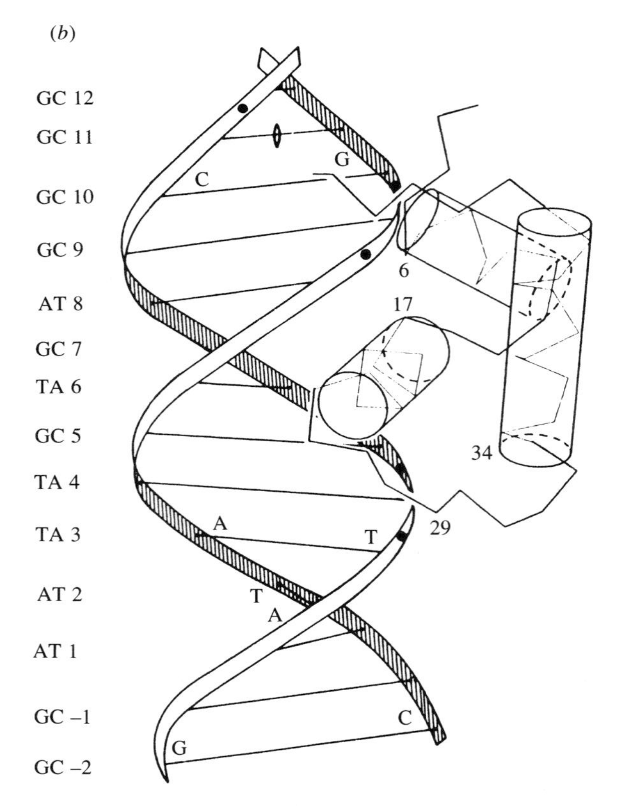

Fig. 15. (*a*) An α-carbon backbone structure of the *N*-terminal 51-residue 'headpiece' of the *E. coli lac* repressor derived from 2D NOE measurements (Kaptain *et al.* 1985). (*b*) Model of *lac* repressor 'headpiece' – *lac* operator complex derived from 2D NOE measurements (reproduced from Boelens *et al.* 1987).

E. coli. The periplasmic-binding proteins are two-domain proteins that have a binding site for sugar in a cleft that lies between the two domains (Pflugrath & Quiocho, 1985; Vyas *et al.* 1988). Mutant *lac* repressor proteins that are defective in inducer-binding (Miller, 1978) have amino-acid changes that are in homologous positions to residues in the arabinose-binding protein that are directly involved in arabinose binding (Vyas *et al.* 1988). Thus the *lac* repressor consists of four elongated core domains that have a structure similar to that of the galactose-binding protein with an inducer-binding site in the middle of each core domain. The DNA-binding domains are located at the ends of each of the four core domains (Fig. 14).

(i) *2D NMR studies of* lac *repressor headpiece*. Nuclear magnetic resonance studies employing 2-dimensional nuclear Overhauser effects (2D NOEs) have been applied to the 51-residue *lac* repressor headpiece and has resulted in a 3D structure for this DNA-binding domain (Kaptain *et al.* 1985). Kaptain and co-workers have directly demonstrated that the region of *lac* repressor showing amino-acid sequence similarity to CAP, *cro* and the other repressors (Matthews *et al.* 1982) does indeed contain a helix–turn–helix structure (Fig. 15*a*).

It has also been possible to identify NOEs between the protein and a DNA substrate (Boelens *et al.* 1987). The approximate model of the 'headpiece'–DNA complex that is consistent with a small set of distances between specific protein and DNA groups place the second helix of the helix–turn–helix in the major groove of B-DNA (Fig. 15*b*) but in a different orientation than that observed in the co-crystal structures of λ*cro*, λ*cI*, CAP, 434 repressor, 434 *cro* and *trp* repressor. Molecular-genetic studies of the binding of mutant repressor proteins to mutant operators have led Müller-Hill and co-workers (Lehming *et al.* 1987) to the same conclusion. In both models amino-acid side chains at the amino end of the 'recognition' helix are interacting with base pairs closer to the dyad axis of the operator than are the side chains in the middle of the helix.

(ii) *DNA crosslinking and looping*. A question that existed for many years was why the *lac* repressor was a tetramer with a potential for two DNA-binding sites. It was clear since the early 1970s that the two-fold symmetry of DNA and most repressor-binding-site sequences would interact optimally with the dimer and that a tetramer should have two DNA-binding sites which indeed the *lac* repressor does. Although pseudo-operator binding sites in the region of the *lac* operator have been known for many years, it has only been recently demonstrated that the *lac* repressor will bind simultaneously to both the *lac* operator and pseudo-operator (Besse *et al.* 1986). It has been shown using band shift gel assays that *lac* repressor will bind to DNA fragments containing two operators as long as the operators are separated by at least six turns of DNA (Kramer *et al.* 1987). It is also necessary that the operators be separated by an integral number of turns so that the repressor may bind on the same side of the DNA in both sites. It may be that the need of *lac* repressor to bind simultaneously to two DNA-binding sites is correlated with the rectangular planar arrangement of subunits. A tetrahedral arrangement of subunits would not orient the two binding sites such that they could easily bind a short loop of DNA.

5.1.2 *Structural studies of the bacterial phage repressors*

Extensive structural, biochemical and molecular-genetic studies have been carried out on repressors from genetic two closely related lytic bacteriophages – the *cI* repressor and *cro* protein from λ and 434 phages (Ptashne, 1986). In both phages these two regulatory proteins have antagonistic roles in phage growth. In the prophage state, when the phage genome is incorporated into that of the bacteria, the *cI* repressor protein is expressed and binds to sites within left and right operators (O_L and O_R) thereby turning off all viral genes that are necessary for lytic growth. Additionally, the *cI* repressor acts as a positive regulator by stimulating the transcription of its own gene. The *cro* protein is expressed during lytic growth and turns off the synthesis of *cI* repressor.

The antagonistic effects of these two repressors are achieved, in part, through their differing affinities for DNA-binding sites with closely similar sequences. These 17 bp binding sites in the 80 bp rightward operator (Fig. 16) are called O_R1, O_R2 and O_R3 while those in the leftward operator are O_L1, O_L2 and O_L3. The sequences in each of these six sites show pseudo-dyad symmetry and are closely similar to each other. *cI* repressor sterically blocks transcription from P_R by binding to O_R1 and O_R2 and stimulates transcription from P_{RM}. *cro*, on the other hand, binds preferentially to O_R3, blocking the further expression of *cI* repressor.

The issues being addressed in structural and molecular-genetic studies of both λ and 434 repressor and *cro* proteins are (1) how do these proteins bind specifically to these similar, but not identical sequences? (2) what accounts for the differing specificity of *cro* and *cI* repressors and (e) how does the *cI* repressor activate transcription by RNA polymerase?

The *cI* repressors are dimers of 26000 molecular-weight polypeptides. Following on the studies of *lac* repressor, it was shown λcI can also be cleaved into two domains by cleavage with papain (Pabo *et al.* 1979). The *N*-terminal fragment contains residues 1–92, binds specifically to the same operator sites as intact repressor and mediates both positive and negative control of transcription (Sauer *et al.* 1979). The C-terminal domain stabilizes the formation of dimers. A similar cleavage of 434 repressor results in a 69 residue *N*-terminal fragment that binds specifically to operator DNA.

High-resolution crystal structure information is now available in all four of these related repressor proteins. The structures of λ*cro* (Anderson *et al.* 1981), 434 *cro* (Wolberger *et al.* 1988) and the DNA-binding fragment of λ repressor (Pabo & Lewis, 1982) have been done as well as DNA complexes with the DNA-binding fragments of λ repressor (Jordan & Pabo, 1988), 434 repressor (Aggarwal *et al.* 1988) and the intact 434 *cro* protein (Wolberger *et al.* 1988). The structures of three of these proteins, 434 *cro*, 434 repressor fragment and λ repressor fragment, are closely similar consisting of five α-helices. The ways in which these three proteins interact with DNA is also similar, though not identical. The structure of λ*cro* protein, however, differs significantly from the others.

In order to compare the repressor operator interactions observed in various systems it is useful to use a common system for numbering the operator base pairs.

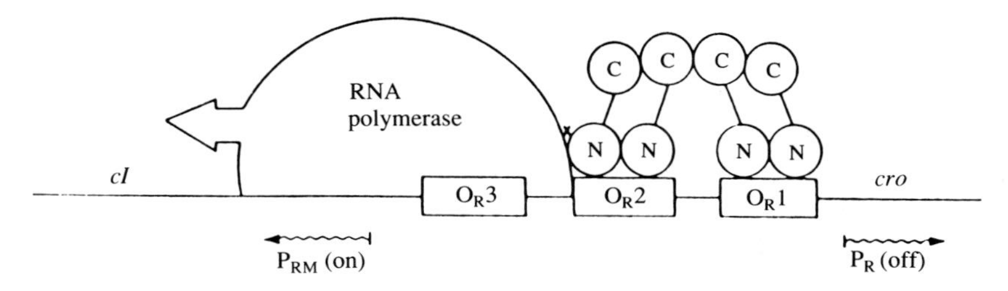

Fig. 16. Schematic drawing of the λ rightward operator (λP_R) with two dimers of intact λ repressor binding to O_R2 and O_R1 and turning off transcription of the *cro* gene and stimulating transcription from P_{RM} (from Lewis *et al.* 1985).

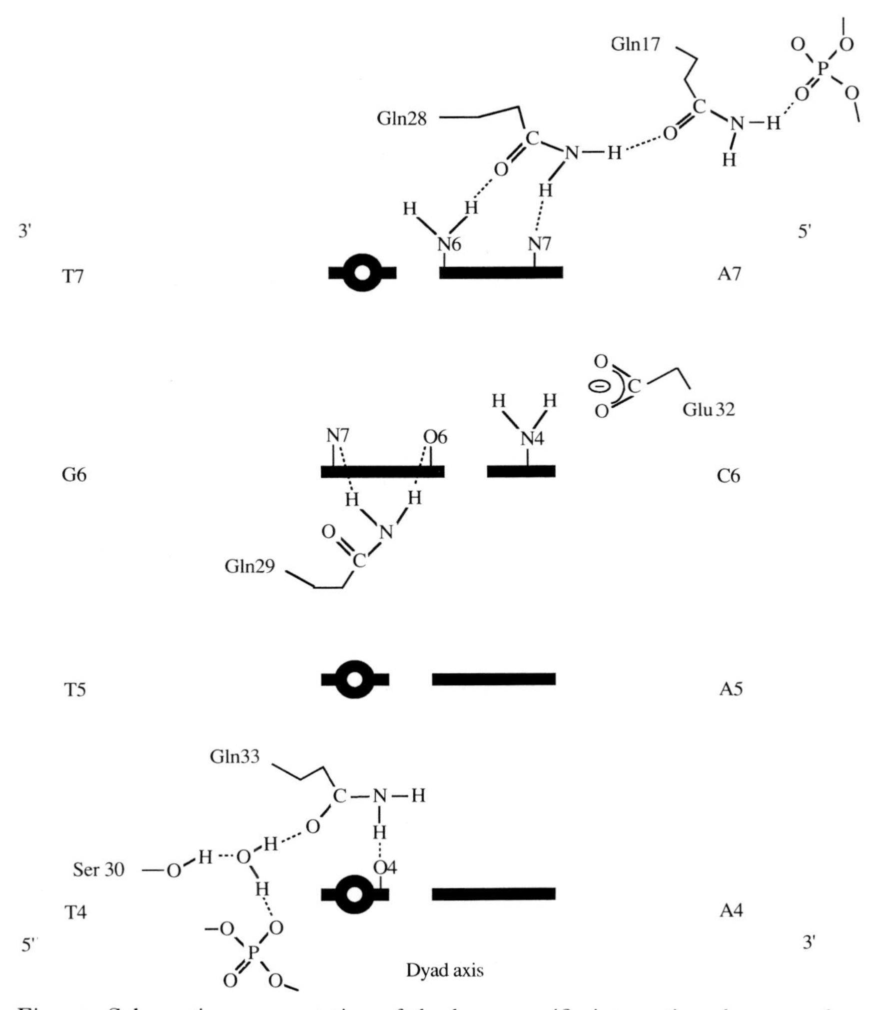

Fig. 17. Schematic representation of the base-specific interactions between the 434 phage repressor and one-half of its operator. Nucleotides are numbered from the central dyad of the operator rather than from the end of a 14 bp operator so that base pair 1 in Aggarwal *et al.* (1988) is base pair 7 in this scheme. The diagram is derived from data presented in Aggarwal *et al.* (1988).

In this review base pairs will be numbered from the dyad axis of symmetry that relates these symmetric sequences. Those operators (e.g. λ) in which the relevant dyad axis passes through a central base pair will be numbered starting with zero for that central base pair. Note that in the primary papers on the repressor–DNA complexes, numbering is either from the dyad axis (TrpR), transcription start (CAP) or semi-arbitrarily defined end of operator (λ*cI* and *cro* and 434 repressor and *cro*).

To facilitate comparisons of direct sequence-specific interactions among the three repressors for which high-resolution co-crystal structures have been published, Figs 17, 18 and 19 schematically compare the base-specific interactions made by these three proteins. Below the structures are schematic summaries of the effects of operator mutations on repressor affinity in each case.

(*a*) *434 Repressor fragment complexed with DNA*

The repressor protein from phage 434 shows a significant degree of sequence homology with the *cI* repressor from phage λ and shares similar functions in the 434 phage. Following on the procedure used for studying the DNA-binding domains of the λ*cI* repressor, an *N*-terminal 69-residue proteolytic fragment of 434 repressor was initially co-crystallized with a 14 bp operator DNA to yield crystals that diffract anisotropically to 3·3 Å in the best direction and 4·5 Å in the worst (Anderson *et al.* 1984). Subsequent co-crystallization of this DNA-binding fragment with a 19 bp duplex DNA containing single nucleotide 5′ overhanging nucleotide at each end yielded crystals diffracting to 2·5 Å, as found earlier by Jordan *et al.* (1985) for the λ repressor fragment complexed with a 19-mer duplex containing single nucleotide 5′ overhang. This size DNA appears important for optimal co-crystallization of both the λ and 434 repressor fragments and 434 *cro*, presumably due in part to the similarity in the size of these proteins and in part to the packing of complexes in the crystal while maintaining end to end stacking of the DNA.

The crystal structure of 434 repressor fragment with the 14-mer (Anderson *et al.* 1987) provided the first direct confirmation of earlier models for the general orientation of the operator DNA on the λ*cI* repressor fragment (Pabo & Lewis, 1982) and provided general support for the models of protein–DNA complexes that were constructed using the protein structure, electrostatic calculations and genetic data on *cro*, CAP and *trp* repressor. Indeed, the second helix of the bihelical motif penetrates into the major groove of B-DNA with a few of its side chains interacting with the exposed edges of base pairs as had been proposed from the earlier model-building studies.

While a number of important features of the specific interaction between 434 repressor and its operator were clear from the medium resolution 14-mer structure, the details of the interaction could only be observed with assurance in the 2·5 Å resolution structure determination of the 20-mer complex (Aggarwal *et al.* 1988). Perhaps the most unanticipated features of this complex are the protein-induced distortions of the DNA structure (Fig. 20). These distortions include an overwinding of the central 6 bp AT-rich region accompanied by a

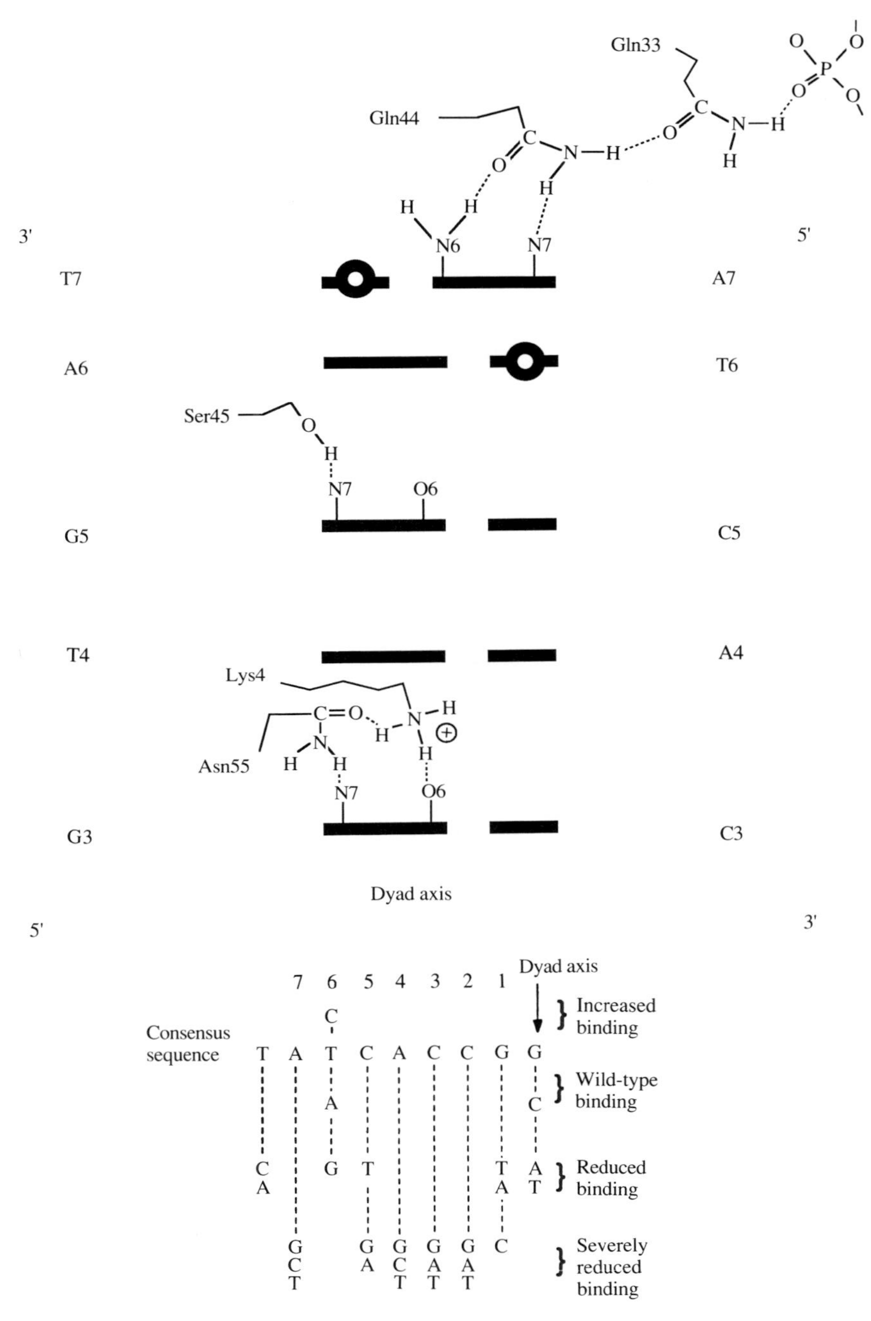

Fig. 18. Schematic representation of the base specific interactions between the λ phage *cI* repressor and one-half of its operator. Nucleotides are numbered from the central dyad of the operator which passes through base pair 0 in this scheme. The diagram is derived from data presented in Pabo & Jordan (1988). Below is a representation of the effects of operator mutations on repressor affinity reproduced from Sauer *et al.* (1990). Note that the sequence-specific interactions observed thus far between repressor and base pairs 7 and 3 are not sufficient to account for the operator mutations. Presumably there are additional interactions/DNA distortions that are not yet seen in the present analysis.

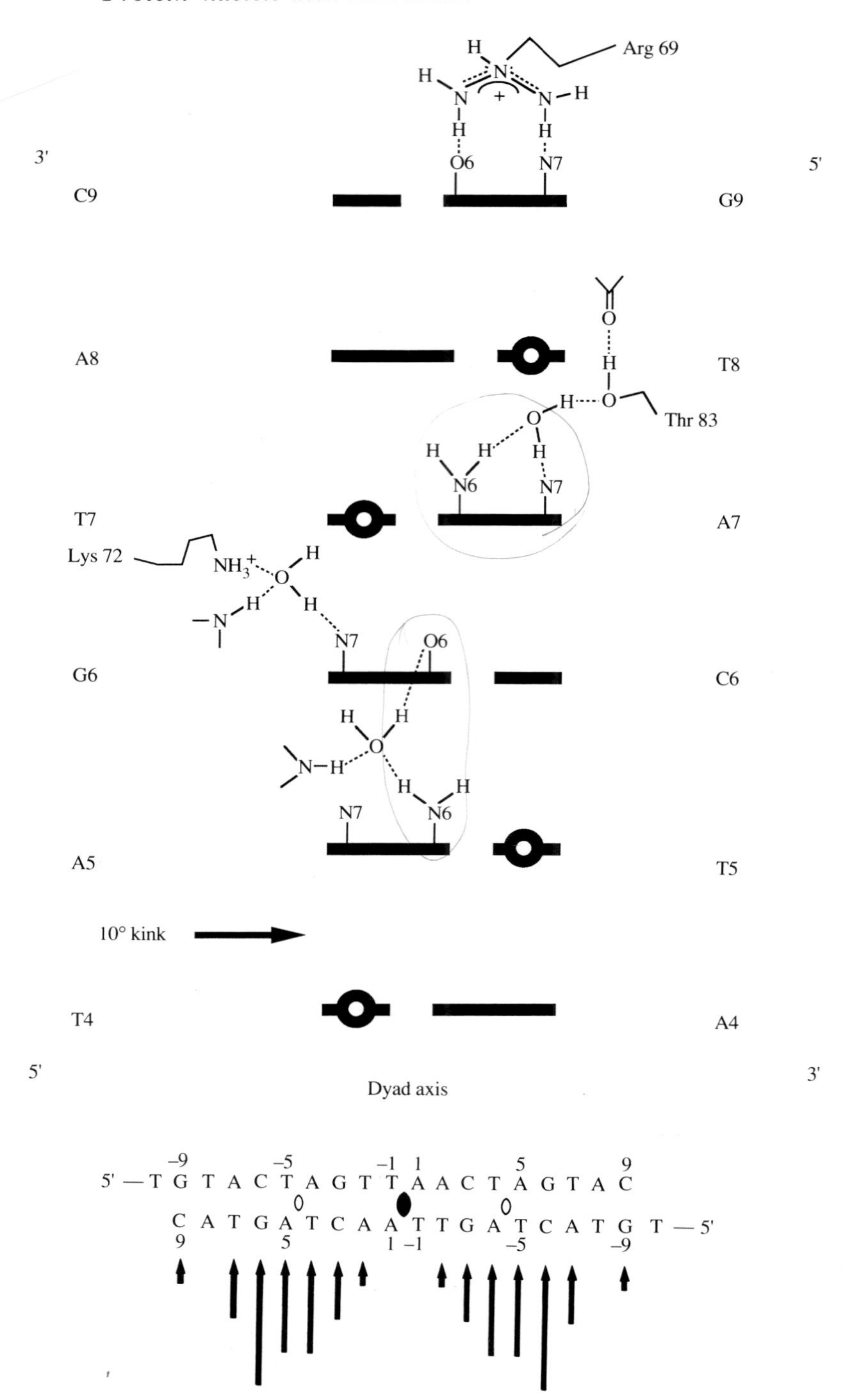

Fig. 19. *Schematic representation of the base-specific trp* repressor–DNA interactions with one half of its operator (Otwinowski *et al.* 1988). Nucleotides are numbered from the central dyad axis. Below is a schematic representation of the affect of operator mutations on repressor affinity (from Otwinowski *et al.* 1988).

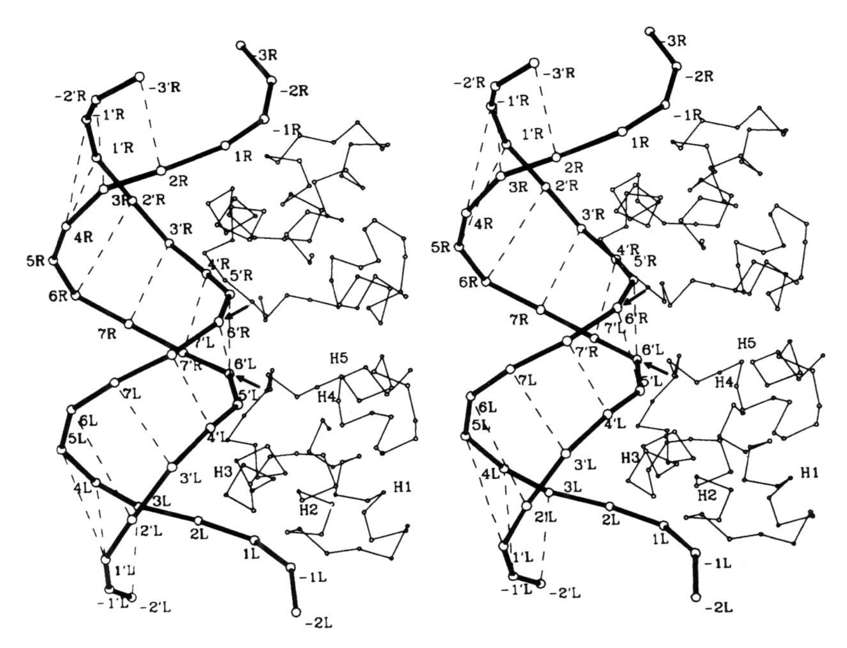

Fig. 20. Stereo view of the phosphate backbone of the 20 bp fragment and the α-carbon backbone of dimeric 434 repressor fragment. The dashed lines connect phosphates across the minor groove. The minor groove is compressed in the middle of the operator (8·8 and 9·2 Å) and then gradually widens at the ends of the operator (13·5–14·0 Å). The turn from helix 3 to helix 4 of R1-69 is close to the phosphate backbone at the centre of the operator. The arrow indicates close approach to phosphate 6′. H1–H5 denote helices 1–5. The numbering of the base pairs is the opposite of that in Fig. 17, i.e. base pair 1 is 7 and base pair 7 is 1 (from Aggarwal *et al.* 1988).

dramatic narrowing of the minor groove to 8·8 Å compared with 11·5 Å for B-DNA (Anderson *et al.* 1987; Aggarwal *et al.* 1988). When examined in detail the centrally located AT base pairs show a high degree of propeller twist (Fig. 21) that allows formation of bifurcated hydrogen bonds between adjacent base pairs of the same type that had been observed in the structure of adenosine-tract DNA (Nelson *et al.* 1987; Coll *et al.* 1987; DiGabriele *et al.* 1989): The O4 thymine forms a hydrogen bond with the N6 of the adenine. Also stabilizing the propeller-twisted bases are water molecules bound in the minor groove. These water molecules are interacting both with the base pairs and the guanidinium group of Arg43 in a manner facilitated by the propeller twisting.

The importance of these distorted nucleotides at the centre of the operator to repressor affinity has been demonstrated by molecular-genetic experiments (Koudelka *et al.* 1987). Replacement of AT by GC base pairs at the dyad axis (particularly 1 and to a lesser extent 2 from the dyad) reduces binding affinity by as much as 50-fold for the intact 434 repressor. Presumably the GC base pairs take up the required distorted structure at high free energy cost, being less prone to high propeller twist.

A second distortion in the DNA involves a bend in the DNA helix axis towards

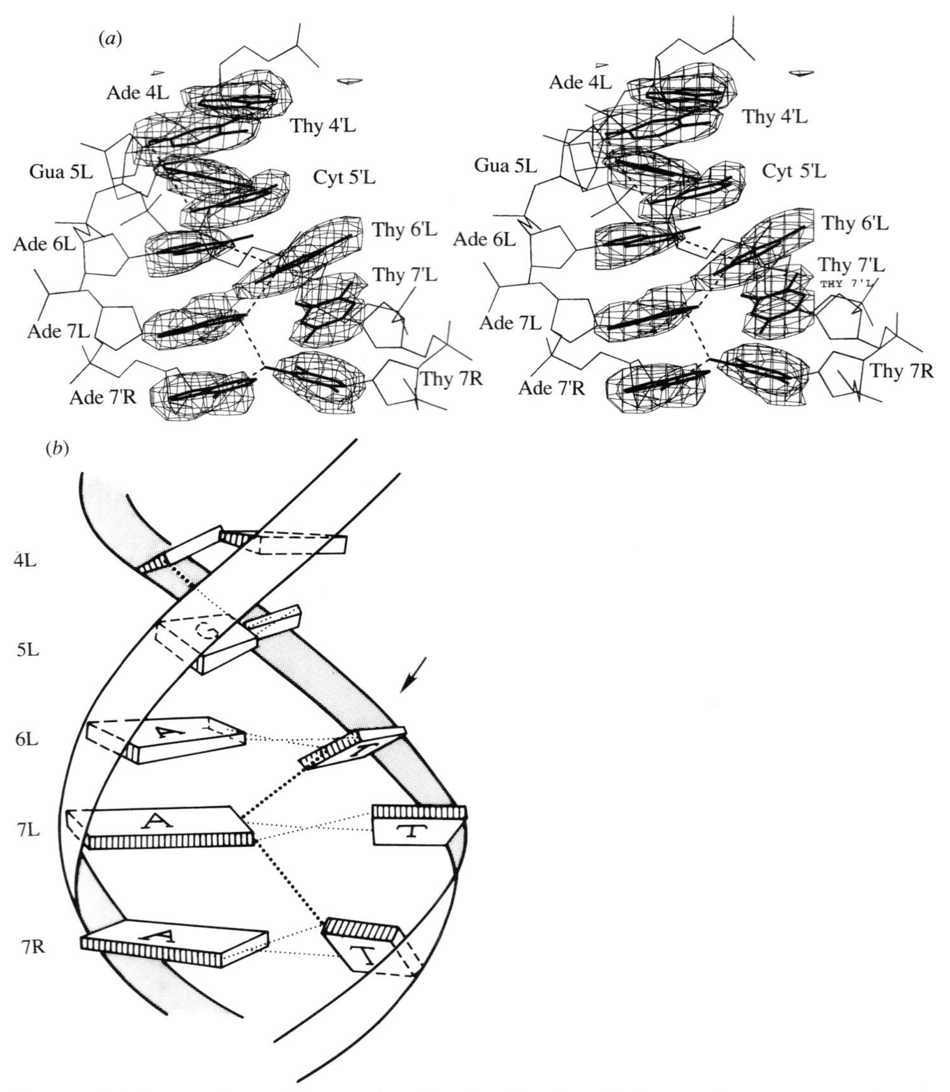

Fig. 21. (*a*) Stereo view of base pairs 4L, 5L, 6L, 7L, 7R (base pairs 4, 3, 2, 1, 1′ in Fig. 17) in the refined structure of the 434 repressor fragment complex. The map is an 'omit map' obtained by leaving out these five base pairs from the calculation of structure factors. The three bifurcated hydrogen bonds shown are made possible by the high propeller twist of the AT base pairs at the dyad axis (7′R and 7L). (*b*) Schematic diagram of base pairs shown in (*a*). Bifurcated hydrogen bonds are shown by heavy dotted lines and Watson–Crick bonds by light dotted lines (from Aggarwal *et al.* 1988). The base pair numbering is opposite that of figure 17, *i.e.* base pair 1 is 7 and 7 is 1.

the protein thereby allowing additional protein–DNA contacts at the extremities of the 20-mer. The overall bend of the DNA seen in the 20-mer complex is substantially larger than that observed in the complex with the 14-mer. A third distortion observed in the complex involves base-pair buckle, the bending of the long axis of a base pair about its short axis. The buckle of base pairs in the 20 bp

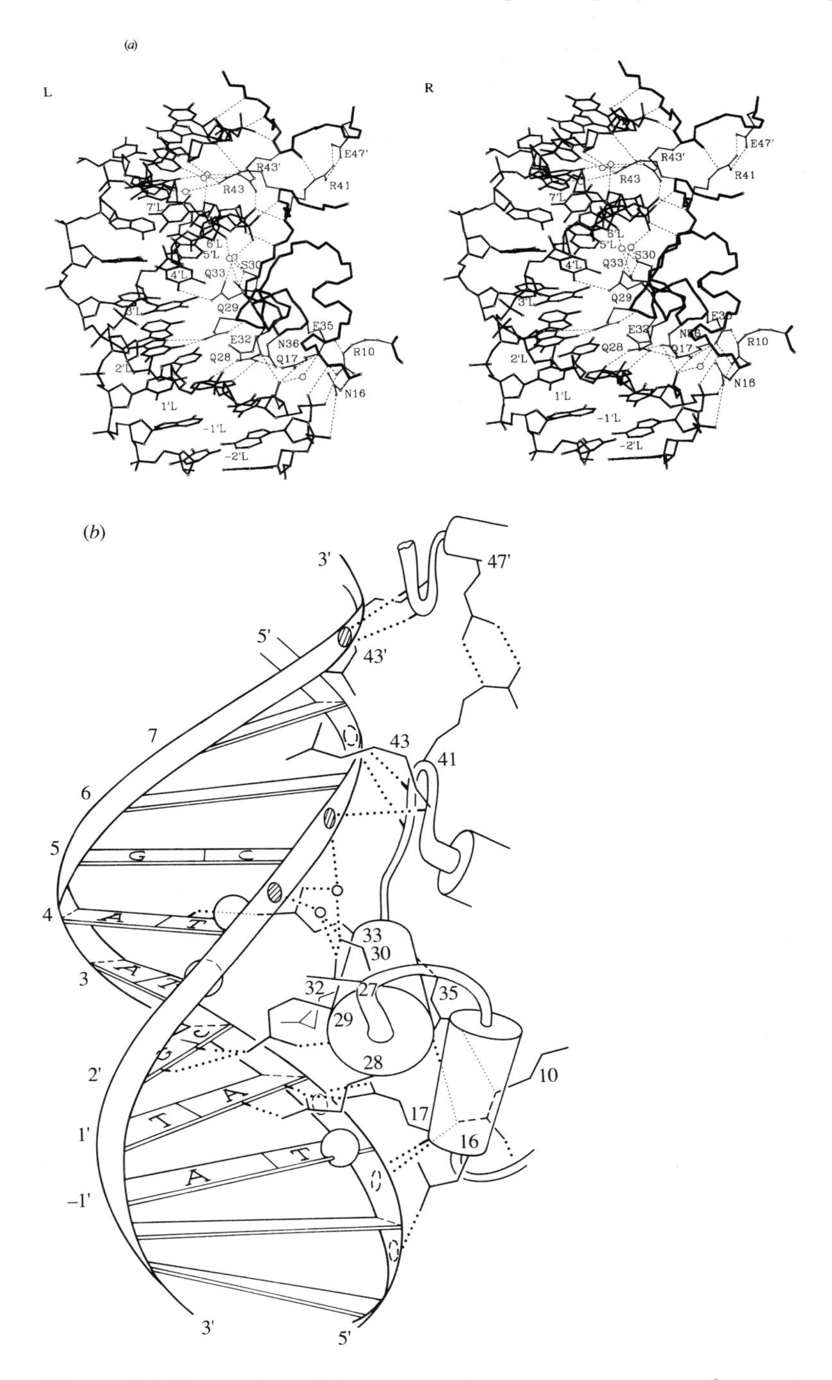

Fig. 22. (*a*) Stereo view of the contacts between 434 repressor fragment and its operator DNA. Hydrogen-bonding interactions (distances up to 3·5 Å) are indicated by dotted lines. (*b*) Diagram of the protein–DNA interface corresponding approximately to the view in (*a*). The base pair numbering is opposite that of figure 17.

fragment varies from $-12{\cdot}9^\circ$ to $19{\cdot}3^\circ$, values that are greater than the $-7{\cdot}0^\circ$ to $5{\cdot}5^\circ$ observed in DNA structures alone (Dickerson & Drew, 1981; Nelson *et al.* 1987).

(i) *Protein–DNA interactions*. Direct interactions are made in the major groove between protein side chains and conserved base pairs that are 4–7 from the DNA dyad axis as well as base pair 8 (Figs 17, 22). The carboxy amide moity of Gln28 makes bidentate hydrogen bonds to the N6 and N7 of adenosine 7 while the aliphatic portion of this side chain makes van der Waals contact with the 5-methyl group of thymine 7. At the next base pair closer to the dyad axis, the ϵ amino group of Gln29 is making a bidentate interaction with the N7 and O6 of guanine 6. While there appear to be no direct hydrogen bonding contracts with the next base pair, the side chains of Thr27 and Gln29 form a hydrophobic pocket to accommodate the methyl group of thymine 5. The side chain of Gln33 once again plays the central role in 'recognition' of the AT base pair that is 4 from the dyad axis. The ϵ amino group of Gln33 is hydrogen bonded to the O4 of thymine 4 while the Oϵ is involved in water-mediated hydrogen bonds to the phosphate backbone in one half of the operator. These direct and sequence specific interactions between the 434 repressor and base pairs 4–7 of its operator are entirely consistent with the observations of Koudelka *et al.* (1987, 1988) and Wharton (1985) that any substitution for ACAA at base pairs 7 to 4 leads to a reduction in the repressor-operator binding constant by more than two orders of magnitude.

(ii) cro *protein and* cI *repressor fragment from phage* λ. Structural and molecular genetic studies of the DNA-binding fragment of *cI* and *cro* repressors from λ phage have been extensively reviewed (Ohlendorf & Matthews, 1983; Takeda *et al.* 1983; Pabo & Sauer, 1984; Ptashne, 1986) so that only recent developments will be described here. These two proteins both bind to the operator control region in λ and regulate transcription from the rightward and leftward promoters. Their DNA sequence specificity for this three-binding-site operator is similar but not identical in important ways.

(*b*) λ*cro*

In solution *cro* is a dimer of identical 66-amino acid polypeptides but crystallizes as a tetramer containing D_2 point group symmetry (Anderson *et al.* 1981). The detailed model of *cro* interacting with its operator DNA proposed by Matthews and colleagues is well known (Ohlendorf *et al.* 1982). The locations of mutations in *cro* that reduce its affinity for operator DNa are consistent with the overall orientation of DNA on *cro* in the model complex (Pabo & Sauer, 1984). The calculated positive electrostatic charge potential of *cro* lies predominantly in the region occupied by DNA in the model. In contrast to CAP, the positive electrostatic charge lies mostly between the two DNA-binding α-helices rather than on the sides of the protein away from the dyad axis (Ohlendorf *et al.* 1983). This may imply that *cro* induces a more modest bend in DNA upon binding than does CAP.

An altered model for the *cro*-operator complex has been proposed on the basis of molecular-genetic studies (Hochschield & Ptashne, 1986*b*). In this model side chains on the recognition helix of λ*cro* that are homologous to λ*cI* and are proposed to interact with DNA are proposed to bind the same base pairs in the λ operator as had been proposed for the *cI* repressor–operator model (Pabo & Lewis, 1982). An earlier comparison of the *cro* and the λ*cI* structures (Ohlendorf *et al.* 1983) had concluded that it was not possible for these two proteins to interact with DNA in the same way unless there is a significant alteration in the orientation of the two subunits of one or the other protein upon binding DNA. However, the contacts between the two subunits of *cro* involve primarily two anti-parallel β-strands giving rise to the possibility that the *cro* subunits do rearrange upon binding DNA. Indeed, the low resolution structure of the complex shows a change in subunit orientation (Brennan & Matthews, private communication 1989).

Several co-crystals of *cro* and various lengths of operator DNA have been grown. Co-crystals with either a 6-mer or a 9-mer duplex showed diffuse scatter between crystal and lattice points at 3·4 Å consistent with the DNA being at least partially disordered in the crystal (Anderson *et al.* 1983). Subsequently, co-crystals that diffract anisotropically to 3·8 Å along one axis and to 6 Å in the perpendicular directions have been grown with a 22 bp DNA fragment (Brennan *et al.* 1986). The structure has been solved to this limited resolution using iodine substituted DNAs and electron density averaging (Brennan & Matthews, private communication). The DNA is bent towards the protein at its ends and α-helix 3 is in the major groove of the DNA as predicted. In this complex the relative orientation of the two subunits has changed from that of the uncomplexed protein, unlike the circumstances with *trp*, λ, 434 *cro* and 434 repressors in which the protein structures remain essentially unchanged upon formation of the complex. In the case of λ*cro*, it is not yet clear how relevant the model-built complex will turn out to be.

(*c*) λcI *repressor fragment*

The crystal structure of an *N*-terminal 92 amino-acid proteolytic fragment of λ repressor was determined by Pabo & Lewis (1982) who constructed a model for its interaction with DNA. Molecular-genetic and DNA-binding studies from the Sauer laboratory showed that the relative orientation of DNa and protein appeared to be correct in this model (Pabo & Sauer, 1984; Nelson & Sauer, 1986). This DNA-binding fragment consists of five α-helices, two of which form the helix–turn–helix structure. The most C-terminal of the helices interacts across a dyad axis with the same α-helix from a second subunit and provides the contacts holding this weak dimer together.

Pabo and co-workers were the first to undertake a systematic search for an appropriate operator DNA to co-crystallize with λ*cI* fragment and obtained co-crystals that diffract to 2·5 Å resolution (Jordan *et al.* 1985). Initial co-crystallizations with a DNA 11-mer and a 17-mer yielded crystals that diffracted poorly and only at low resolution. Realizing that the end to end DNA–DNA interactions were important in the crystal packing, Pabo and colleagues prepared a series of duplex operator DNAs that varied in length between 18 and 22 bp. They also produced duplex DNAs that were either blunt end or containing a single base or two base, single stranded 5′ overhang. While many of these DNAs would indeed co-crystallize with the *λcI* fragment protein, most crystals were of poor quality. The best co-crystals were obtained using a 19 bp operator DNA which contained a single complementary 5′ nucleotide overhang on each end.

(i) *Structure of λ repressor–DNA complex.* The crystal structure of a complex between the DNA-binding domain of λ repressor and a 20 bp λ operator site has recently been determined at 2·5 Å resolution and refined to a crystallographic *R*-factor of 0·24 (Jordan & Pabo, 1988). The overall arrangement of the complex is similar to that predicted from the earlier model-building studies in which the dimeric protein fragment was fitted against straight B-form DNA (Pabo & Lewis, 1982). There are, however, important interactions observed in the crystal structure that were not predicted from either model-building or genetic studies. As is the case in most of the other DNA co-crystals, the duplex DNA is stacked end to end in the crystal and the existence of a 5′ overhanging T on one strand and a complementary 5′ overhanging A on the other strand of the duplex apparently facilitates this end to end stacking in the crystal.

Unlike the DNA complexes with *Eco*R I, 434 repressor, *trp* repressor and CAP, the duplex DNA in these crystals is not observed to be strikingly kinked. The largest deviations from linear B-DNA occur at the ends of the operator site which bend slightly toward the repressor. The DNA is not, however, uniform and the helical twist angle from one base pair to the next varies from 47° to 21° and the degree of propeller twists observed between the two halves of the base pair varies from a high of 21° to a low of 2·9°. The extent to which this variation in the DNA structure is inherent in the DNA sequence and the extent to which it is induced upon binding to the protein is unknown as the structure of the free DNA has not been determined.

λ repressor contacts exposed edges of bases in the major groove and the phosphate oxygens of the sugar phosphate backbone (Fig. 23). Much of the sequence specificity is thought to occur through direct hydrogen bonds between side chains of Gln44 and Ser45 that are at the amino end of the recognition helix and by Asn55 that occurs in a loop just after the recognition helix and Lys4 that occurs in the amino terminal arm.

As predicted in the model built complex, Gln44 forms 2 hydrogen bonds with the adenine of base pair 6 from the dyad axis (Figs 18, 24). The amide NH_2 of the glutamine side chain donates a hydrogen bond to the N7 of adenine while the carbonyl oxygen accepts a hydrogen bond from the N6, as also observed in the 434 co-crystals with DNA. Not predicted in the earlier model building, however, is an interaction between Gln44 and the side chain of Gln33 which also hydrogen bonds to the oxygen of a phosphate. Once again, a similar interaction is observed in the 434R co-crystals (Aggarwal *et al.* 1988).

As anticipated from earlier models, the λ-hydroxl of Ser45 hydrogen bonds to the N7 of guanine 4. Without additional hydrogen bonds, however, this interaction is not sequence specific since it could presumably hydrogen bond to the N7 of an adenine at that position as well. the second sequence-specific hydrogen-bonding interaction involves Asp55 and Lys4 (Fig. 24). Asn55 hydrogen bonds to the N7 of adenine while the ϵ-amino group of Lys4 forms a

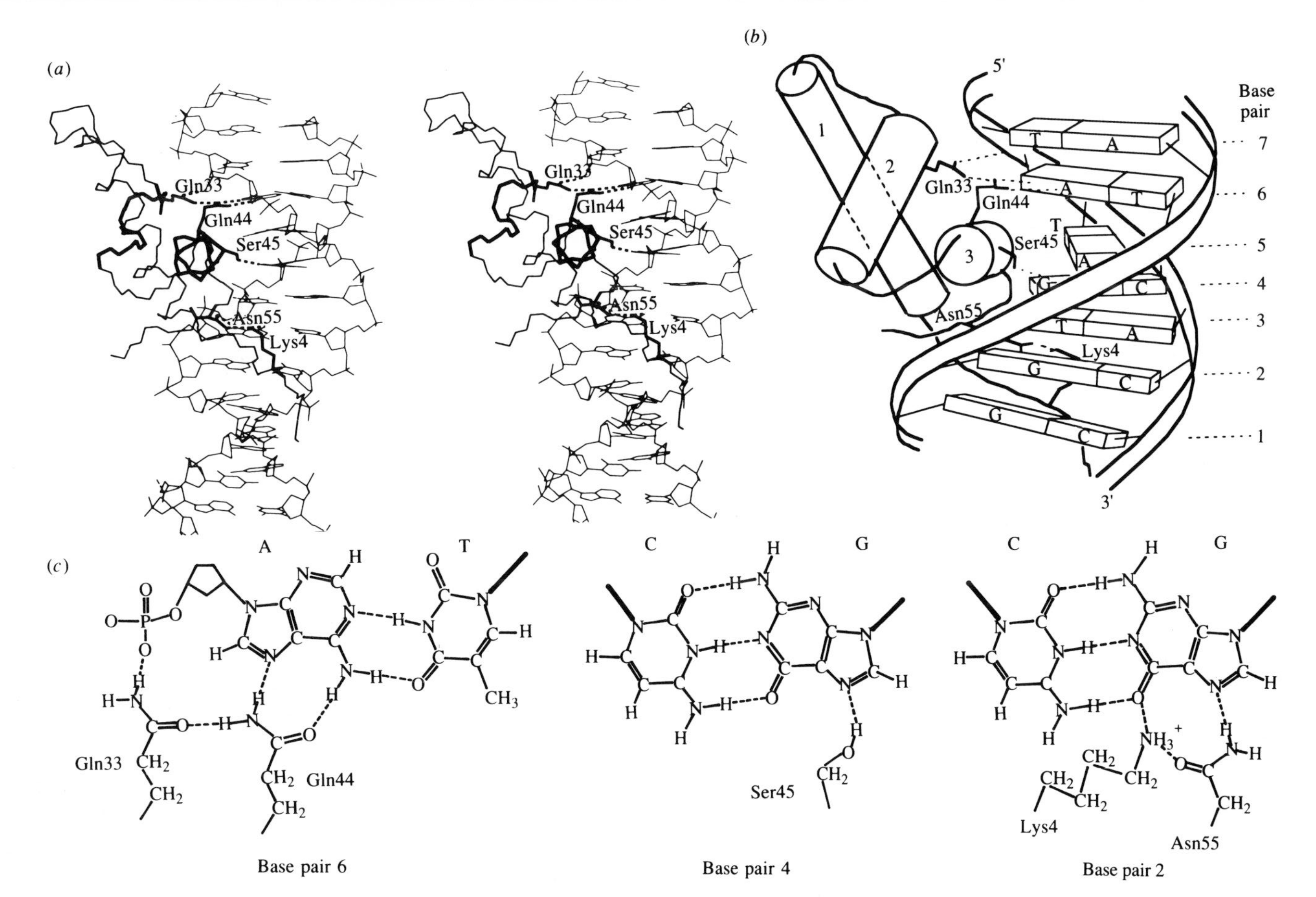

Fig. 24. (*a*) Stereo diagram of the λ repressor fragment–DNA complex showing the side chains that interact with base pairs in the major groove. The backbone of the protein is shown for residues 1–58, and the NH_2-terminal arm begins in the lower right corner of this figure. Critical side chains, and the backbone of helices 2 and 3, are emphasized with bold lines. (*b*) Sketch, in the same orientation as (*a*), showing hydrogen bonds in the consensus half-site. In the upper half of the figure, the major groove is readily visible for base pairs 1–3; in the lower half, the minor groove is closest to the viewer. (*c*) Sketches summarizing how side chains hydrogen-bond to base pairs 2, 4 and 6 in the consensus half-site. The solid lines show the connection to the sugar phosphate backbone (from Jordan & Pabo, 1988).

bridging hydrogen bond between the amide carbonyl of Asn55 and the O6 of guanine.

Thus far it appears that the structural basis for λ repressor 'recognizing' two base pairs per half-operator site has been delineated. This results from Gln44 H-bonding to adenine 6 bp from the DNA dyad axis and Asp55 and Lys4 H-bonding to adenine 4 bp from the dyad axis. As genetic data on operator mutations show that additional base pairs in the operator are recognized (Bensen *et al.* 1988), there must be additional sequence-specific interactions between the repressor and DNA. Some of these interactions appear to involve the amino terminal 'arm' of the repressor that was shown by biochemical means to wrap around the protein (Pabo *et al.* 1982).

5.1.3 E. coli *Trp repressor*

The crystal structure of the tryptophan repressor from *E. coli* confirms some previously derived principles and establishes some new ones. The *trp* repressor is a dimer of identical 108 amino-acid subunits. In the presence of L-tryptophan it binds to the *trp* operator blocking transcription of the operon by RNA polymerase. Unlike the *lac* repressor which binds to DNA and represses in the absence of the allosteric inducer ligand, the *trp* repressor binds and represses only as a complex with the allosteric ligand tryptophan. Since the *trp* operon produces enzymes for a biosynthetic pathway, the accumulation of the product of the pathway, tryptophan, is required to shut down enzyme biosynthesis. In the *lac* operon the presence of the catabolite galactose (and therefore the inducer allolactose) in the medium is a signal for derepression of the degradative enzymes of the lactose operon while the absence of galactose is the signal to turn off synthesis of these enzymes.

The *trp* repressor has been crystallized without tryptophan (*apo trp* repressor), as a complex with tryptophan (in two crystal forms) and as a complex with tryptophan and a DNA operator fragment. Structures of all four of these crystals have been solved and refined at very high resolution (1·8–2·4 Å) making available on this system complete structural information on all relevant ligation states of this allosterically regulated protein (Schevitz *et al.* 1985; Zhang *et al.* 1987; Otwinowski *et al.* 1988).

(i) *Repressor structure.* Upon examination of the structure of the *trp*–repressor complex with *trp* at 1·8 Å resolution, Sigler and colleagues have divided the dimeric *trp* repressor into three structural domains (Schevitz *et al.* 1985). The first 60 amino acids of each subunit form three α-helices that are intimately intertwined about the dyad axis (Fig. 25). This portion is termed the 'hard core' since its structure remains invariant in the four solved crystal structures. The remainder of the sequence forms a helix–turn–helix. The α-carbon coordinates of the helix–turn–helix of *trp* repressor superimposed on the corresponding motifs of *cro*, CAP and *λcI* repressor with RMS differences of 0·7, 1·0 and 1·0 Å respectively (Zhang *et al.* 1987).

The bound tryptophan molecule is almost completely buried between the D and

Fig. 25. Schematic drawing of the *trp* repressor dimer in stereo viewed down the molecular dyad axis relating the two intertwined subunits. The α-helices are represented as tubes lettered from the amino terminus. The co-repressor L-tryptophan is drawn in skeletal form (from Schevitz *et al.* 1985).

E α-helices and the 'hard core' portion of the protein. The indole ring of tryptophan occupies a position that is normally filled by a bulky hydrophobic side chain in the homologous repressor proteins.

The structure of the *trp apo* repressor has been refined at 1·8 Å resolution and shows a conformation difference from the repressor–tryptophan complex that accounts for the *apo* repressor's inability to bind DNA specifically (Zhang *et al.* 1987). In the absence of bound tryptophan the helix–turn–helix motif of *trp* repressor collapses onto the 'hard core' changing the relative orientation of the two protruding DNA-binding helices of the dimer (Fig. 26). The fact that both structures have been refined at 1·8 Å resolution has allowed a very precise description of the change in hydrogen-bonding pattern that occurs upon tryptophan binding (Fig. 27). Since the *apo* repressor does not have a structure that is complementary to B-DNA and presumably does not bind specifically, the reduction by 10^3 in the *trp apo* repressor affinity for specific-sequence DNA probably reflects the equilibrium constant between the active and inactive conformations of the protein in the absence of tryptophan. Mutations in the repressor that serve to favour the active conformation in the absence of tryptophan will increase the affinity of the *apo* repressor for DNA. One such mutation that has been examined (Zhang *et al.* 1987) has changed a glycine to a valine. The valine side chain of the mutant repressor prevents formation of the collapsed conformation shown by the wild-type *apo* repressor and therefore stabilizes the DNA-binding conformation. This mutation in *trp* repressor is quite analogous to the CAP* mutations that allow CAP to bind DNA specifically in the absence of cAMP.

(ii) *Structure of* trp *co-repressor complexed with DNA*. The crystal structure of the

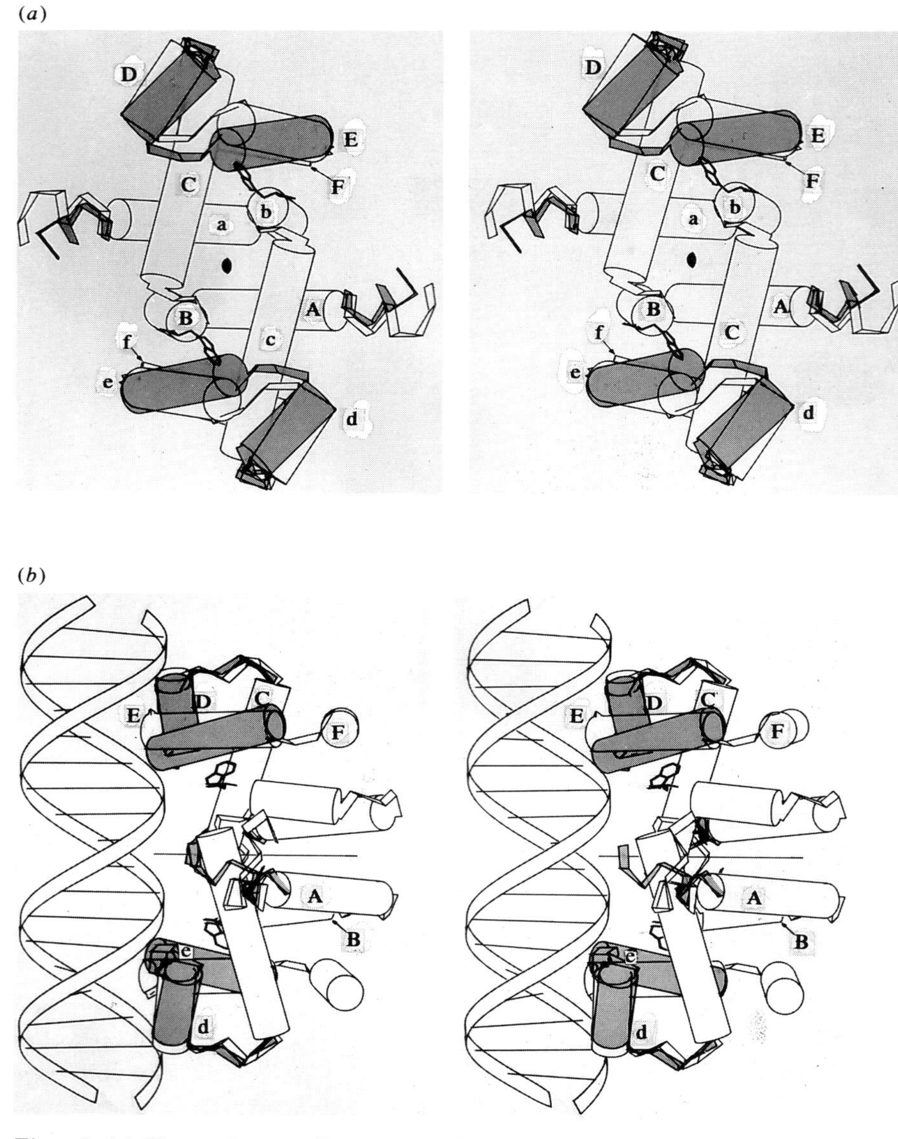

Fig. 26. (*a*) Change in overall structure of the *trp* repressor upon binding L-tryptophan. The most notable change is in the positions of the helix–turn–helix motifs (D and E) relative to the 'hard core'. The *apo* repressor changes are shown in grey (from Zhang *et al.* 1987). (*b*) The two superimposed repressor structures docked onto DNA. Changes in the position of the helix–turn–helix alters the complementarity between the repressor and DNA structures (from Zhang *et al.* 1987).

dimeric *trp* co-repressor complexed with a symmetrical 18 bp duplex with overhanging 5′T residues has been established at 2·4 Å resolution (Otwinowski *et al.* 1988). The reported structure has been refined to a crystallographic *R* factor of 0·25 and recently data have been extended to 1·9 Å resolution and the structure further refined (Sigler, personal communication, 1989). At this resolution and

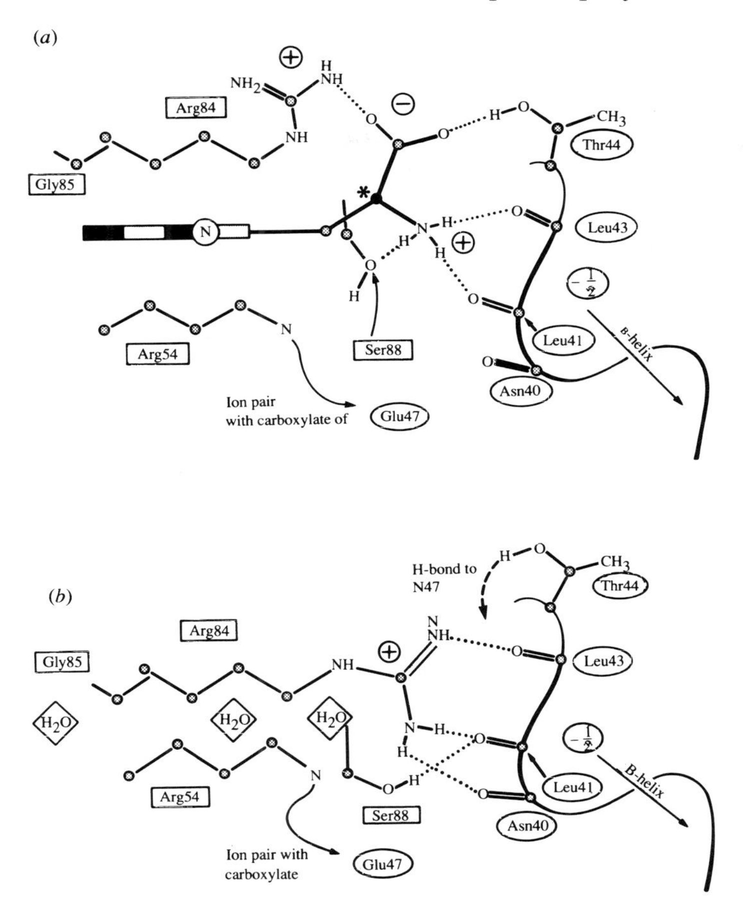

Fig. 27. Schematic drawing of the detailed changes in hydrogen-bonding patterns that occurs upon binding of tryptophan to TrpR (from Zhang *et al.* 1987). (*a*) H-bonding interactions made to tryptophan. (*b*) H-bonding interactions made in absence of tryptophan.

level of refinement it is possible to not only clearly delineate the contacts between protein and nucleic acid, but also to identify the positions of water molecules that appear to participate in the interaction between the two macromolecules.

While the structure of the complex shows that the amino end of the helix–turn–helix is penetrating the major groove of the operator DNA as anticipated by analogy by other repressor–operator systems, there appears to be only one direct hydrogen-bonding contact between the *trp* repressor and the bases, the guanidinium group of Arg69 interacting with G9 (Figs 19, 28). This was not completely unanticipated (Schevitz *et al.* 1985) since the first two amino acids of the recognition helix in position to contact DNA are isoleucine and alanine in the *trp* repressor. Instead of direct hydrogen-bonding contacts between the protein and the bases, *trp* repressor appears to have a number of important water-mediated contacts between the protein and the base pairs (Otwinowski *et al.* 1988). It appears that most of the effect of operator mutations on its affinity for repressor can be rationalized in terms of the crystal structure of the complex. Mutations in base pairs 3–7 from the dyad axis have the largest effect on repressor affinity while mutations and base pairs 9 and 2 have small effects. It appears that base pairs 4 and 5 may be important to allow the DNA to undergo the 12° bend or kink between them that is required for the outer part of the DNA to make suitable contact with the repressor. The AT base pair at position 5 undergoes a 12° roll into the major groove narrowing slightly the major groove. From studies of the bending of DNA around nucleosomes Drew & Travers (1984) have concluded that AT base pairs favour kinking in the direction of closing the major groove while GC base pairs favour kinks in the direction of the minor groove.

Recognition of A7 appears to be achieved via a water-mediated hydrogen bond to the N6 and N7 of adenine. This water-mediated hydrogen bond is sequence specific because the interaction with the hydroxyl of Thr83 requires that water donates a hydrogen bond. This allows water to be a hydrogen-bond donor and acceptor but would not allow the water to donate a hydrogen bond to G. While the 5-methyl group of T might disrupt the making of even one hydrogen bond, it is less clear what would happen in the case of C at this position. The network of water-mediated hydrogen bonds with base pairs 5 and 6 also suggests a high degree of sequence selectivity.

A question concerning the correct DNA fragment to use for the *trp* operator-binding site has been raised. Since the *trp* operator shows dyad symmetry in the sequence every 4 bp, a binding site displaced by 4 bp from the one used by Otwinowski *et al.* (1988) has been suggested (Phillips *et al.* 1989; Müller-Hill, personal communication). Further, binding and structural experiments on additional DNA fragments may prove enlightening on this issue.

5.1.4 E. coli *Met repressor*

The *E. coli* Met repressor (MetR) represses transcription of its own gene and those coding for enzymes involved in methionine and S-adenosyl-methionine (SAM). Methionine is the precursor of SAM which functions as the main intracellular methyl donor. SAM is the corepressor whose binding to the MetR is required for this dimer of 12000 molecular-weight subunits to bind to its operator DNA.

The structure of the MetR (Rafferty *et al.* 1989) shows two highly intertwined monomers, reminiscent of the *trp* repressor (Fig. 29). In this case, however, a β-strand from each subunit is intertwined to form a two-stranded anti-parallel β-ribbon that in the operator–repressor co-crystal structure is seen to fit into the major groove of B-DNA. There are in addition three α-helices that form part of the SAM-binding pocket. There is in this structure no helix–turn–helix motif, thus requiring a different mode of interaction with DNA from the other bacterial repressors of known structure.

The crystal structures of the Met repressor with and without bound SAM have been determined from crystals grown under different conditions and possessing different unit cell and space group parameters. The surprising observation is that these two liganded states of this allosterically controlled repressor have almost

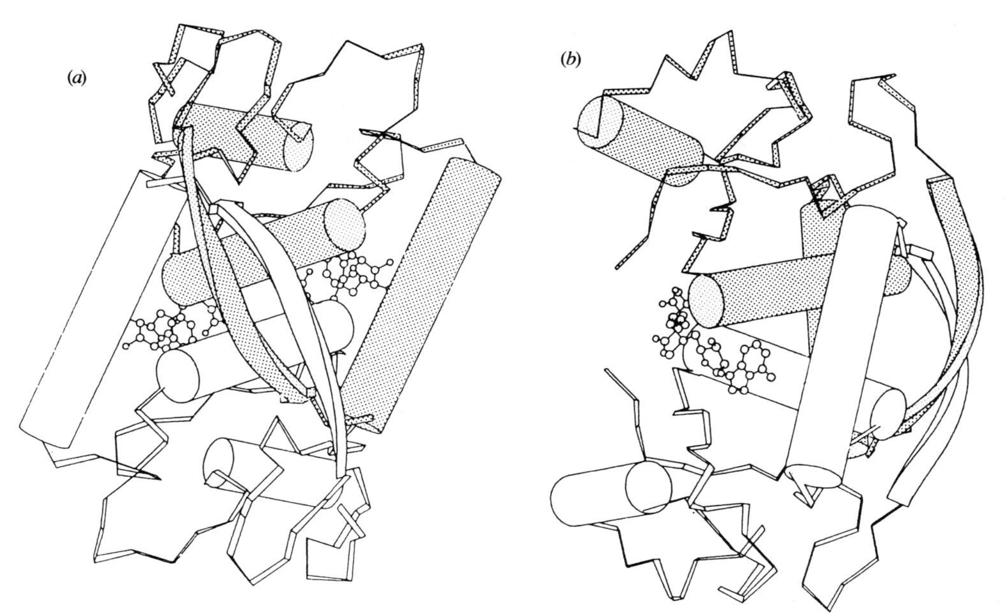

Fig. 29. Schematic drawing of the *Met* repressor dimer with one subunit shaded. The co-repressor SAM is shown in 'ball and stick' representation. (*a*) View down the molecular dyad axis. (*b*) View with molecular dyad axis horizontal. The protruding anti-parallel β-ribbon that interacts with DNA is on the right (from Rafferty *et al.* 1989).

identically the same structure. The r.m.s. derivation for the fitting of corresponding α-carbon atoms of *apo* repressor to the SAM–repressor complex (excluding flexible loops, was 0.43 Å compared with 0.34 Å for the fitting of the non-crystallographically related monomers of the two dimers. Since these two nearly identical structures apparently do not explain the different affinity of *apo*- and SAM-bound repressors for operator DNA, one must consider the possibility that the crystallization conditions have put both ligation states of the protein into either the R or the T state.

In trying to model the complex between the MetR and DNA, Rafferty *et al.* (1989) docked the repressor onto the DNA in such a way as to put the two A α-helices of each subunit into successive major grooves of B-DNA. Subsequent determination of the repressor–DNA structure (Rafferty *et al.* 1989, note added in proof) showed that the protein was in fact interacting with the DNA utilizing its opposite face and the anti-parallel β-ribbon formed by the two subunits. It would be of interest to know whether calculations of the positive and negative electrostatic charge potential (e.g. Warwicker *et al.* 1987) would have identified the correct surface for docking with the DNA.

5.1.5 *Zinc-containing DNA-binding domains (zinc 'fingers') TFIIIA*

Transcription factor IIIA (TFIIIA) from *Xenopus* is a 40000 molecular-weight protein that binds specifically to 5S RNA forming a 7S complex. It also binds to the gene for 5S RNA to positively regulate its transcription. Its binding to DNA protects approximately 60 bp from DNAse I, a distance of nearly 200 Å indicating that TFIIIA must be a substantially elongated protein (Bogenhagen *et al.* 1980).

Klug and co-workers (Miller *et al.* 1985) discovered that the portion of the TFIIIA sequence known from deletion analysis to confer the sequence-specific binding contained a series of nine repeats of homologous 30-amino-acid residue units. Among the conserved amino acids in these 30-amino-acid repeating units were two histidines and two cysteines proposed by these workers to bind the 7–9 zincs found to be associated with TFIIIA. EXAFS studies have since confirmed the interaction of zinc with histidine and cysteine ligands (Diakun *et al.* 1986). The model proposed for TFIIIA interaction with DNA consists of a series of these units interacting with 5 bp repeating sequences in the DNA. The term zinc 'finger' was coined to describe the interaction of these units with DNA.

A creative model (Fig. 5) of a zinc finger was constructed by analogy to other zinc-binding proteins whose crystal structure is known. Jeremy Berg (1988) suggested that the two cysteines in this sequence form a structure and interact with a zinc in the same fashion as observed for the CTP-binding domain of aspartic transcarbanylase and an iron–sulphur protein. He also found that in thermolysin histidines lying on a α-helix form two ligands with the zinc. Putting these two facts together he was able to construct a compact model of a zinc-binding domain that contains two strands of anti-parallel β-sheet and a single α-helix with a zinc atom lying between these two secondary structural elements bound by the cysteines on the β-strands and histidines on the α-helix. This model also placed the other conserved hydrophobic residues in the interior of this small domain. Such a model suggested that the zinc fingers might interact in the major groove of B-DNA via an α-helix (Berg, 1988) in a fashion not so dissimilar from that observed for the repressors and *Eco*R I.

More recently the structures of two different zinc 'finger' domains have been determined by 2D NMR methods (Parraga *et al.* 1988; Lee *et al.* 1989). The more detailed model derived by Lee *et al.* (1989) confirms the overall aspects of the Berg model, though some small differences exist. This may constitute the first time that a globular protein structure has been correctly predicted from its amino-acid sequence (depending perhaps on what one means by 'correct', 'predict' and 'first').

5.2 *Restriction endonuclease* – E. coli Eco*R I*

The DNA-recognition sites and their recognition by restriction endonucleases differ from the DNA-binding sites of gene regulatory proteins in several ways. (1) The DNA sequence recognized by the restriction endonuclease is usually centred at the dyad axis of the DNA, whereas the sequences primarily recognized by repressors tend to lie 5–6 bp on either side of the central dyad axis. (2) The restriction sites recognized by the enzymes tend to be 4–8 bp in length and showing a strict dyad symmetry in the DNA sequence, whereas the DNA sequences recognized by repressors extend over 14–20 bp in length (or more) and show only partial dyad symmetry. (3) The restriction endonucleases exhibit a much greater degree of sequence specificity in the enzymatic reaction than is exhibited in the binding of repressors to DNA. A single base-pair change in a critical operator sequence usually reduces the affinity for the repressor by 10- to 100-fold, whereas a single base-pair change in the recognition site of an endonuclease essentially eliminates all enzymatic activity.

(i) Eco*R I endonuclease complexed with DNA*. A most significant advance in our understanding of the structural aspects of protein–nucleic acid interaction came from the determination of the crystal structure of *E. coli Eco*R I restriction endonuclease complexed with a 12 bp DNA duplex containing a single nucleotide overhang (McClarin *et al.* 1986). *Eco*R I functions as a dimer of two identical 31 000 molecular-weight subunits and catalyses the cleavage of a double-stranded sequence d(GAATTC). The enzyme hydrolyzes the phosphodiester bond between the guanylic and adenylic acid residues resulting in a 5′ phosphate. The enzyme requires Mg^{2+} for phosphodiester bond hydrolysis but binds to a cognate hexonucleotide in the absence of Mg^{2+} with a dissociation constant on the order of 10^{-11} M^{-1}.

*Eco*R I was initially crystallized with the same dodecamer sequence, d(CGCGAATTCGCG) containing the recognition sequence, that was crystallized and solved by Dickerson & Drew (1981) (Dickerson, 1983). To prevent hydrolysis of the substrate, crystals are grown in the presence of EDTA. These initial crystals were thin and diffracted only moderately well. However, co-crystals grown with a dodecamer sequence containing a 5′ single-stranded thymidine overhang produced crystals that were significantly more suitable for high-resolution crystallographic analysis. As in the case of the co-crystals of 434 repressor and λ repressor with DNA, these crystals form with DNA duplex packed end to end throughout the crystal. Apparently, small alterations at the ends can significantly alter the crystal packing and the consequent quality of the crystals.

The crystal structure was determined a 3 Å resolution using a single, platinum, heavy atom derivative (McClarin *et al.* 1986). While a single, heavy atom derivative, even using anomalous scattering, is not usually sufficient to determine the macromolecular structure, application of solvent flattening methods developed by Wang (1987) enhanced the quality of electron density map sufficiently to allow its interpretation. With these methods [which are essentially the same as developed by Bricogne (1976) for multiple subunits in the asymmetric unit] one makes use of the knowledge that the solvent must be flat to improve the phasing of the observed diffraction amplitudes and consequently the electron density map.

(ii) *Protein-induced DNA kinking*. Since the structure of the DNA in the absence of the protein has been determined (Dickerson & Drew, 1981), it is possible for Rosenberg and colleagues to say with some degree of assurance that the binding of *Eco*R I to the DNA produces at least two kinds of very significant distortions, or kinks, in the DNA (Frederick *et al.* 1984; McClarin *et al.* 1986). The first kind of kink which they term a 'neo-1' kink occurs at the central dyad

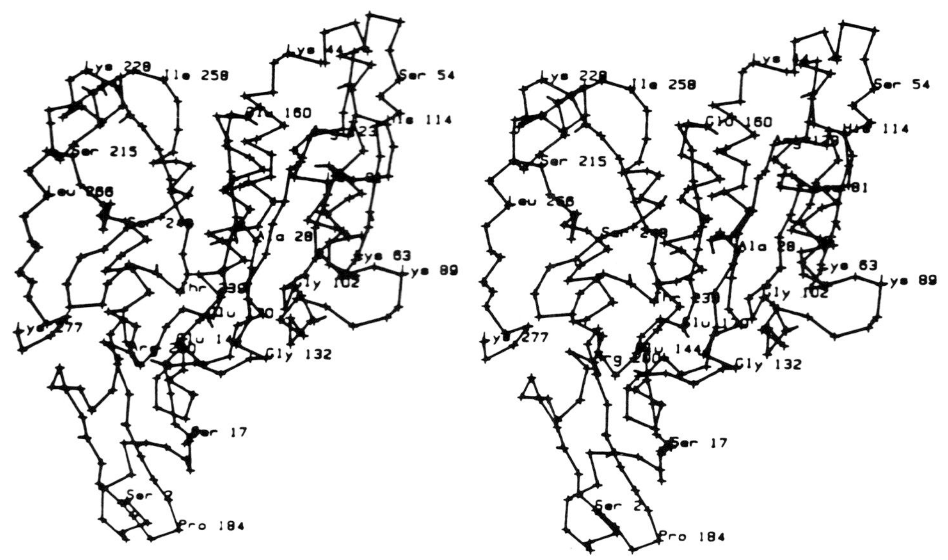

Fig. 30. A stereo drawing of the α-carbon backbone of *Eco*R I (from McClarin *et al.* 1986), but see note added in proof p. 75.

axis and is characterized by an unwinding of the DNA by approximately 25°, the same value obtained by Sung-hou Kim from solution measurements of the unwinding angle (Kim *et al.* 1984). This unwinding of the top 6 bp relative to the bottom 6 bp by 25° results in a widening of the major groove at the dyad by about 3·5 Å. The widening allows the two α-helices from each subunit of the dimer to fit (end on) into the major groove. Further, the realignment of base pairs produced by the neo-1 kink creates sites for multiple hydrogen bonds with the protein not present in the undistorted DNA.

A second kind of kink observed 3 bp from the dyad axis on both sides of the dyad gives rise to a bend between the DNA axes of the DNA on either side of the kink. This kink, termed a 'neo-2' kink, results from a roll of adjacent base pairs into the minor groove. Although the precise angle of the bend cannot be determined in the unrefined structure with great precision, it is estimated to be a bend of between 20° and 40° (McClarin *et al.* 1986).

(iii) *Protein structure.* The *Eco*R I protein is characterized as an α-β protein since it contains a 5-stranded β-sheet surrounded on either side by α-helices (Fig. 30). As is characteristic of enzymes containing this structural motif, the active site of the exonuclease lies at the C-terminus of this parallel β-sheet and forms a catalytic cleft. The amino ends of the α-helices point towards the DNA thus placing the positive end of the helix dipole in contact with the DNA. The *N*-terminus of the protein forms an arm that partially wraps around the DNA.

(iv) *DNA recognition.* A major portion of the sequence specificity exhibited by this enzyme appears to be achieved through an array of twelve hydrogen bond donors and acceptors from protein side chains that are complementary to the donors and acceptors presented by the exposed edges of the base pairs in the hexonucleotide recognition sequence. Two arginines (Arg145α, Arg200β) and one

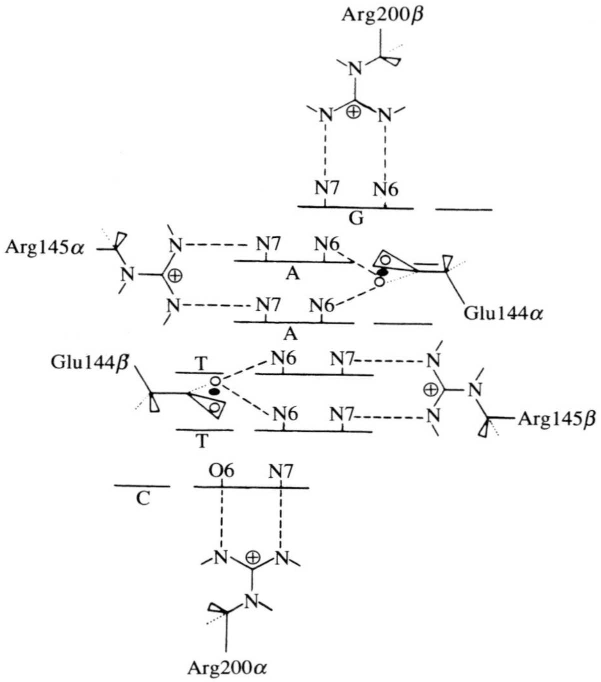

Fig. 31. A diagrammatic representation of the hydrogen-bond interactions between *Eco*R I and its cleavage site that render it sequence selective (from McClarin *et al.* 1986), but see note added in proof p. 280.

glutamic acid (Glu144β) each make two hydrogen bonds to the DNA bases in both subunits of the dimer (Fig. 31). These three side chains are able to penetrate the deep major groove by emanating from the ends of two α-helices that protrude from the protein into the DNA major groove. Thus, a bundle of four parallel α-helices, two from each dimer, is pushed into the major groove for the purpose of direct DNA-base sequence recognition (Fig. 32).

Each of the three side chains is making two hydrogen bonds with bases. Arg200 is hydrogen bonded to the N7 and O6 of guanine in a manner suggested earlier on the basis of model building by Seeman *et al.* (1976). The other two side chain interactions serve to cross-link the adenine of two adjacent base-pairs. Arg145 from one subunit donates hydrogen bonds to the N7 of two adjacent adenines while Glu144 from the other subunit accepts hydrogen bonds from the N6 amino groups of the same two adjacent adenines that bind the arginine.

While this array of hydrogen bond donors and acceptors may account for much of the sequence selectivity of this enzyme, it may not be sufficient to account entirely for the very low rate of cleavage of incorrect sequences. Of particular significance in this regard is the observation of Modrich and co-workers who find that replacement of Glu111 with glycine produces a mutant protein retaining full DNA binding specificity but showing no cleavage activity under physiological conditions. Glu111 is not near the DNA and cannot directly participate in the formation of either the recognition or cleavage sites.

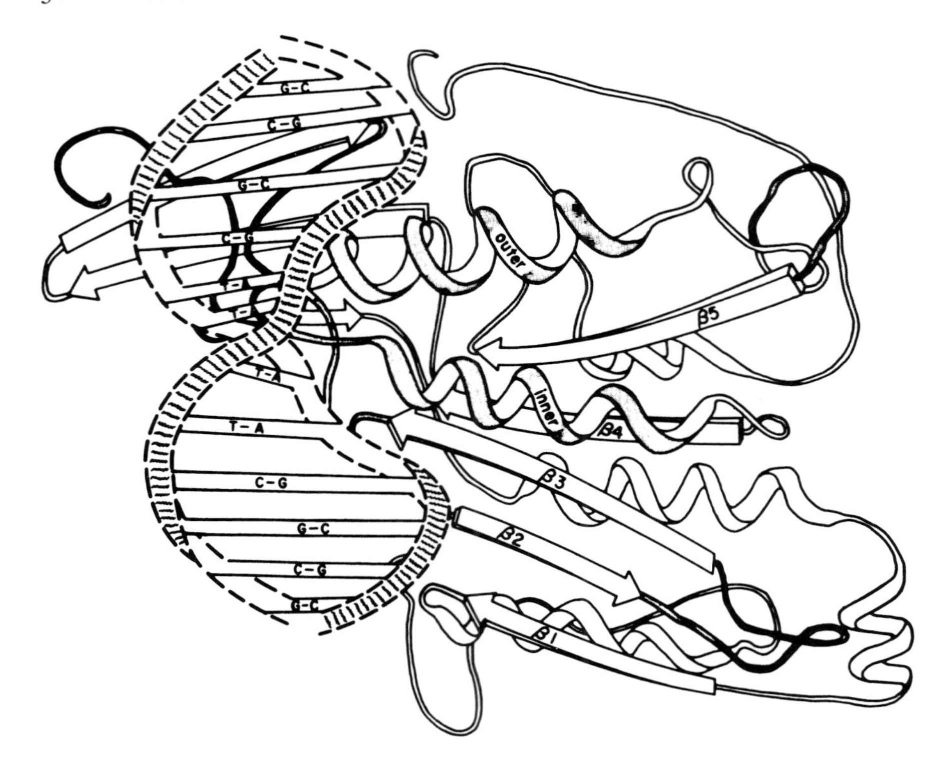

Fig. 32. A schematic representation of *Eco*R I fitting into DNA. The two α-helices that present the side chains which interact directly with the bases are shaded and called 'inner' and 'outer' (from McClarin *et al.* 1986).

Both Modrich and Rosenberg and co-workers have arrived at an allosteric activation or induced fit hypothesis to account for the higher specificity of cleavage as compared with binding. Rosenberg suggests that a catalytically active form of the enzyme is only formed once the requisite sequence-specific DNA protein interactions have formed (McClarin *et al.* 1986).

It has been pointed out (Fersht, 1985), however, that induced fit cannot by itself increase enzyme specificity since the free energy required to convert the enzyme from the inactive to the active form is subtracted from the free energy of binding of the specific as compared with the non-specific substrate; no selectivity is gained as compared with an enzyme already in the appropriate conformation. However, if a reaction is broken down into two or more steps, each one of which shows the substrate specificity, then the overall specificity of the reaction is the product of the selectivities of each step. As this is the way that very high levels of substrate or sequence accuracies are achieved in the reactions catalysed by DNA polymerases, aminoacyl tRNA synthetases and ribosomes, it may be the mechanism employed by restriction enzymes as well.

5.3 *Site-specific recombination – γδ Resolvase*

The γδ resolvase is a 20500 Da polypeptide encoded by the transposon γδ, a member of the Tn3 family of prokaryotic transposable elements (for review see Hatfull & Grindley, 1988). The primary function of resolvase is to catalyse an intra-molecular site-specific recombination between two identically oriented copies of a transposon γδ. The natural substrate is a co-integrate molecule, an intermediate formed during γδ transposition, that consists of donor and target replicons fused together with a copy of γδ at each junction. In addition, resolvase as a transcriptional repressor that regulates expression of the divergently transcribed *TnpA* (transposase) and *TnpR* (resolvase) genes.

In carrying out the site-specific recombination reaction resolvase exhibits the activities characteristic of a repressor, restriction enzyme, topoisomerase and ligase. Resolvase binds to three dyad symmetric sites that span approximately 120 bp of the intergenic region (termed the *res* site), all three of which are required for maximal recombination efficiency (Fig. 33). Recombination occurs near the dyad axis of site 1. As an intermediate in the recombination reaction, Ser10 becomes covalently linked to the 5′ phosphate at the cleavage site.

The 183-residue resolvase protein has been cleaved into the fragments using chymotrypsin (Abdel Meguid *et al.* 1984). The 140-residue amino terminal fragment mediates the protein–protein interactions and has the active site for catalysis. Footprinting analysis shows that the 43-residue C-terminal peptide binds in a sequence-specific manner to each of the six half-sites in the *res* region (Fig. 33). Steitz & Weber (1985) found that 21 of the 43-residue C-terminal peptide show significant sequence homology to the helix–turn–helix of CAP, *cro*, λ*cI* and other bacterial repressors (Abdel-Meguid *et al.* 1984). Thus, γδ resolvase (as well as the family of homologous recombinases and invertases) recognizes its specific target sequence in a manner that is very similar to that of the bacterial regulatory proteins discussed above.

Both the intact protein and the catalytic domain form hexagonal crystals that are anisotropically disordered diffracting to 3 Å along the C* axis and 4 Å along the a* axis (Weber *et al.* 1982*c*; Abdel-Meguid *et al.* 1986). The diffraction intensities from the native protein and fragment crystals are very similar showing that the catalytic domain makes the protein contacts in the crystal and that the DNA-binding domain is at least partially disordered relative to the large domain. That the DNA-binding domain is disordered is now confirmed by the crystal structure of the hexagonal form (Rice & Steitz, unpublished). The catalytic domain also crystallizes in an orthorhombic form that diffracts to 2·4 Å resolution (Abdel-Meguid *et al.* 1986).

(i) Structure of the resolvase catalytic domain. The structure of the catalytic domain has been solved and partially refined at 2·5 Å resolution (Hatfull *et al.* 1989; Sanderson *et al.*, 1989, unpublished). Heavy atom derivatives were made for this structure determination using site directed mutagenesis to introduce cysteine residues capable of specifically binding mercury ions (Hatfull *et al.* 1989). The structure reveals three subunits in the crystallographic asymmetric unit related by

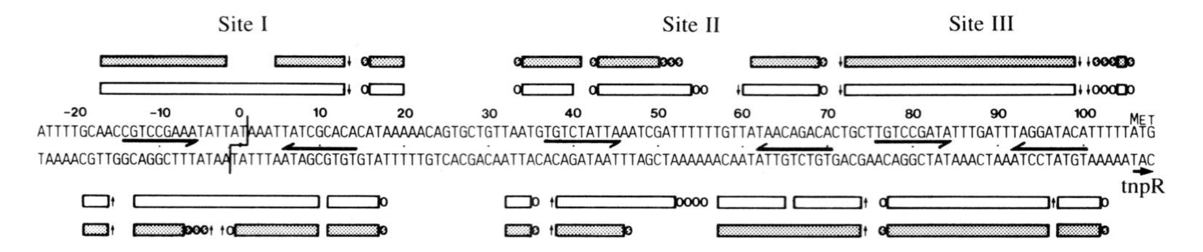

Fig. 33. The *res* site of $\gamma\delta$ showing regions protected from digestion by DNase I by the binding of intact resolvase (open bars) and the C-terminal 43-amino acids of resolvase (shaded bars). Arrows mark the positions of enhanced cleavage. The recombination crossover point is indicated in site I. The nine base sequences that show sequence similarity to each other are indicated by horizontal lines with half arrowheads indicating orientation (from Abdel-Meguid *et al.* 1984).

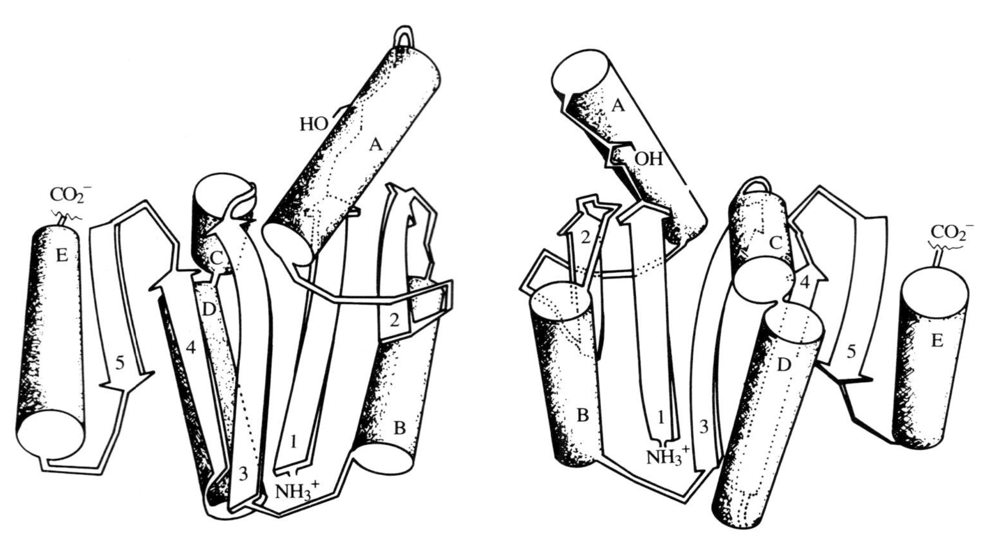

Fig. 34. A schematic drawing of the catalytic domain (residues 1–120) of $\gamma\delta$ resolvase. The two subunits that are packed about the dyad axis in the crystal that places the two Ser10 residues (–OH) closest are shown (Rice *et al.*, unpublished).

two non-crystallographic 2-fold axes. Each subunit consists of a five-stranded β-sheet with α-helices on either side. Using Ser10 as a marker of the active site, the recombinase catalytic site then lies at the carboxy end of a parallel β-sheet, as occurs with *Eco*R I. Considering the subunits in pairs, there is only one pair of subunits in the asymmetric unit of the crystal that places Ser10 close to a dyad axis (Fig. 34). This is an expected feature of the protein dimer that binds to the cleavage site and becomes covalently crosslinked via Ser10 to the 5′ phosphate of the DNA. The two 5′ phosphates that constitute the cleavage site lie 13 Å apart across the minor groove of standard B-DNA. Attempts to dock DNA onto this dimer and place the 5′ phosphate of the DNA adjacent to the hydroxyl of Ser10 as must occur in the intermediate in the reaction requires pulling the duplex DNA apart and twisting the DNA. It is not yet established, however, whether this dimer is in fact relevant in the recombination reaction, since the contact surface is small and highly variable in sequence among the large number of homologous resolvases. Nevertheless, side chains of mutant proteins devoid of recombinase but not DNA-binding activity lie on the protein surface adjacent to Ser10 and at the subunit interface between this dimer and a dyad axis related dimer, suggesting an important role for the tetramer (and this dimer).

The structure of co-crystals of resolvase with DNA presently under study will be required to understand the structural basis of this recombinase.

6. SEQUENCE-INDEPENDENT DNA-BINDING PROTEINS

6.1 *Klenow fragment of* E. coli *DNA polymerase I*

DNA polymerase I from *E. coli* (Pol I), unlike many other replication enzymes, is active as a single subunit (molecular weight 103 000). The enzyme has three enzymatic activities: a DNA polymerase, a 3′-5′ exonuclease that edits out mismatched terminal nucleotides and a 5′-3′ exonuclease that removes DNA ahead of the growing point of a DNA chain. Limited proteolysis of Pol I removes the 35000 Da amino terminal domain that retains its 3′-5′ exonuclease activity (Klenow & Henningson, 1970; Brutlag *et al.* 1969). The remaining 68000 Da large fragment (Klenow fragment) retains the polymerization in editing 3′-5′ exonuclease activities.

The crystal structure of the Klenow fragment (Ollis *et al.* 1985) shows that the 605 amino-acid polypeptide is folded into two distinct structural domains of approximately 200 and 400 amino acids (Fig. 35). A combination of crystallographic and molecular-genetic data have proven that the smaller amino terminal domain catalyses the 3′-5′ exonuclease activity while the larger C-terminal domain contains the active site for the polymerase reaction (Ollis *et al.* 1985; Freemont *et al.* 1986; Derbyshire *et al.* 1988; Steitz *et al.* 1987; Beese *et al.*, unpublished). The carboxy terminal domain (residues 515–928) has been overexpressed and purified and shows polymerase activity but no measurable exonuclease activity (Freemont *et al.* 1986). Rush and Konigsberg have crosslinked the dNTP analogue, 8-azido dATP, to Tyr766 in the large domain (Joyce *et al.* 1986). Deoxynucleoside triphosphate is seen to bind within the cleft of the large domain in a 3 Å resolution difference electron density map (Beese & Steitz, unpublished). Finally, numerous mutants within the proposed polymerase active site have reduced polymerase activity but essentially unchanged exonuclease activities (Polesky & Joyce, personal communication).

Evidence that the small domain catalyses the exonuclease reaction was suggested initially by its ability to bind dNMP, since dNMP was known (Que *et al.* 1978) to inhibit the 3′-5′ exonuclease reaction. More recently, site-directed mutagenesis of residues within the dNMP binding region produce protein that is completely devoid of 3′-5′ exonuclease activity but retains polymerase activity (Derbyshire *et al.* 1988).

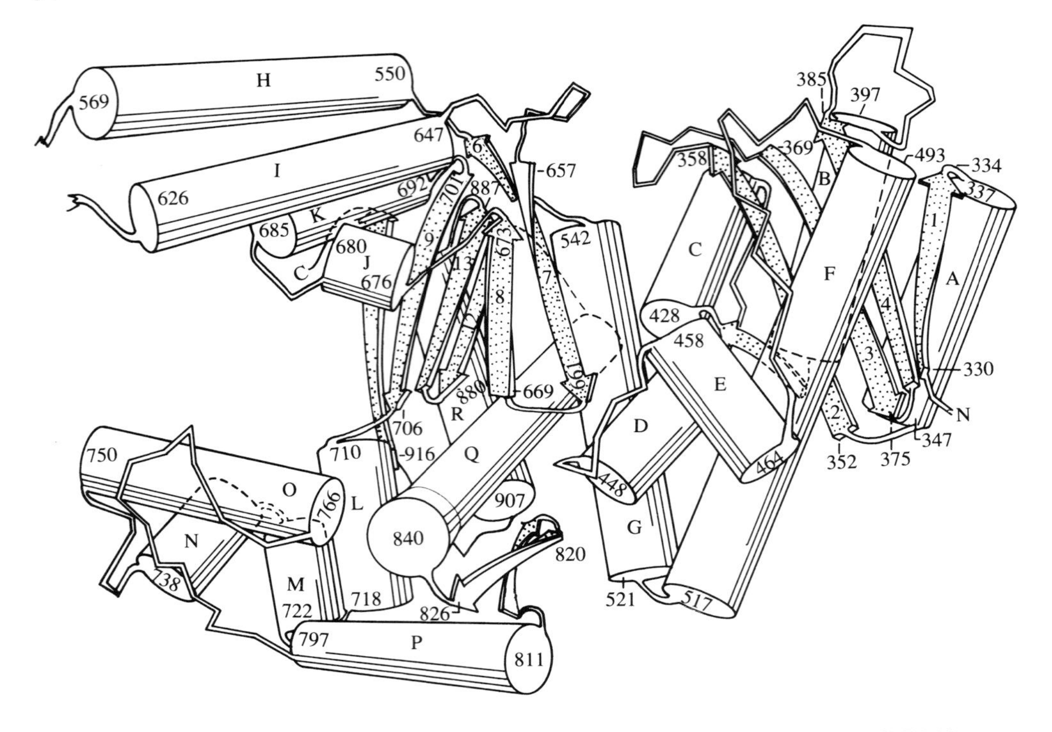

Fig. 35. Schematic drawing of the large proteolytic (Klenow) fragment of *E. coli* DNA polymerase I. The domain catalysing the 3′-5′ exonuclease reaction includes residues 324–520 while the polymerase domain includes residues 521 to the C-terminus and contains a large cleft (from Ollis *et al.* 1985).

(i) *Exonuclease active site.* Co-crystallization of Klenow fragment devoid of exonuclease activity with duplex DNA produces a co-crystal with single-stranded DNA bound at the exonuclease active site (Freemont *et al.* 1988; Steitz *et al.* 1987). The refined structure of the deoxynucleoside monophosphate complex (Beese and Steitz, unpublished) shows the product deoxynucleoside monophosphate bound at the exonuclease active site with its α-phosphate interacting directly with one of the two metal ions. The more tightly bound metal ion is liganded to the enzyme through the carboxylate groups of Leu357, Asp501 and Asp355. A second metal ion is 4·3 Å away and interacting directly with Asp355.

A single-stranded tetranucleotide is seen to bind to the exonuclease active site with its 3′ terminal nucleotide in exactly the same position as that of the nucleoside monophosphate (Fig. 36). The few interactions observed with bases are hydrophobic and non-sequence specific. They include an interaction of Phe473 and Leu361 with a 3′ terminal base. Leu361 is wedged between the 3′ terminal base and the penultimate base in a manner quite analogous to that observed in the binding of glutaminyl-tRNA synthetase to glutamine-tRNA (Rould *et al.* 1989). There are numerous hydrogen-bonding interactions with the sugar–phosphate backbone and the bases are facing away from the protein, consistent with the sequence-independent character of this DNA-binding site.

Based on the crystal structure of these complexes at the exonuclease active site, a specific proposal was made for an enzymatic mechanism of the exonuclease reaction utilizing only the two metal ions (Freemont *et al.* 1988). One metal (A) is proposed to act as a Lewis acid to produce the attacking hydroxide ion while the other (B) is hypothesized to stabilize the pentacovalent transition state and leaving oxyanion. There are essentially no amino-acid side chains in the vicinity of the phosphodiester bond to be hydrolysed that looked to be suitable candidates for catalytic residues. Derbyshire & Joyce (personal communication, 1990) have changed all the residues in the active site region by site-directed mutagenesis. Only those residues involved with binding the metal ions have a catastrophic effect on the exonuclease activity consistent with the hypothesis that the metals themselves are vital for catalysis.

How can a binding site that shows no sequence preference in binding single-stranded DNA edit out mismatched base pairs in the newly synthesized DNA? Since the polymerase-active site binds duplex DNA and the *exo* site binds single-stranded DNA, it was proposed that the lower melting point of the mismatched DNA favoured its forming single-stranded DNA and binding to the exonuclease active site (Steitz *et al.* 1987; Joyce & Steitz, 1987; Freemont *et al.* 1988; Brutlag & Kornberg, 1972). In other words, there is a sequence dependence to the propensity of a duplex DNA to assume the structure required for binding to the exonuclease active site.

(ii) *Polymerase active site.* The extent of structural information available on the polymerase active site interaction with substrates is substantially less than on the exonuclease active site. Although the Klenow fragment has been co-crystallized with a duplex DNA substrate under low ionic strength conditions in which the enzyme is active, the current crystal structure of this putative complex does not show the presence of DNA (Friedman & Steitz, unpublished).

A model has been constructed that places the duplex product of DNA synthesis in the large cleft (Ollis *et al.* 1985). Amino-acid residues that are highly conserved among the prokaryotic DNA polymerases whose sequence is known are concentrated in this cleft and particularly in the vicinity of the model built 3′ terminus of the substrate DNA. A binary complex between deoxynucleoside triphosphate and the Klenow fragment has been formed in crystals grown initially at high salt and transferred to low ionic strength. The dNTP is found to lie in the large cleft (Beese *et al.* unpublished).

6.2 *Bovine pancreatic DNase I*

Although bovine pancreatic DNase I does not show the highly specific sequence-dependent cleavage shown by restriction enzymes, such as *Eco*R I, its cutting rates do vary along a DNA sequence (Scheffler *et al.* 1968; Lomonossoff *et al.* 1981; Drew & Travers, 1984). It has been suggested that this sequence preference arises from the enzyme recognizing sequence-dependent variations in the DNA

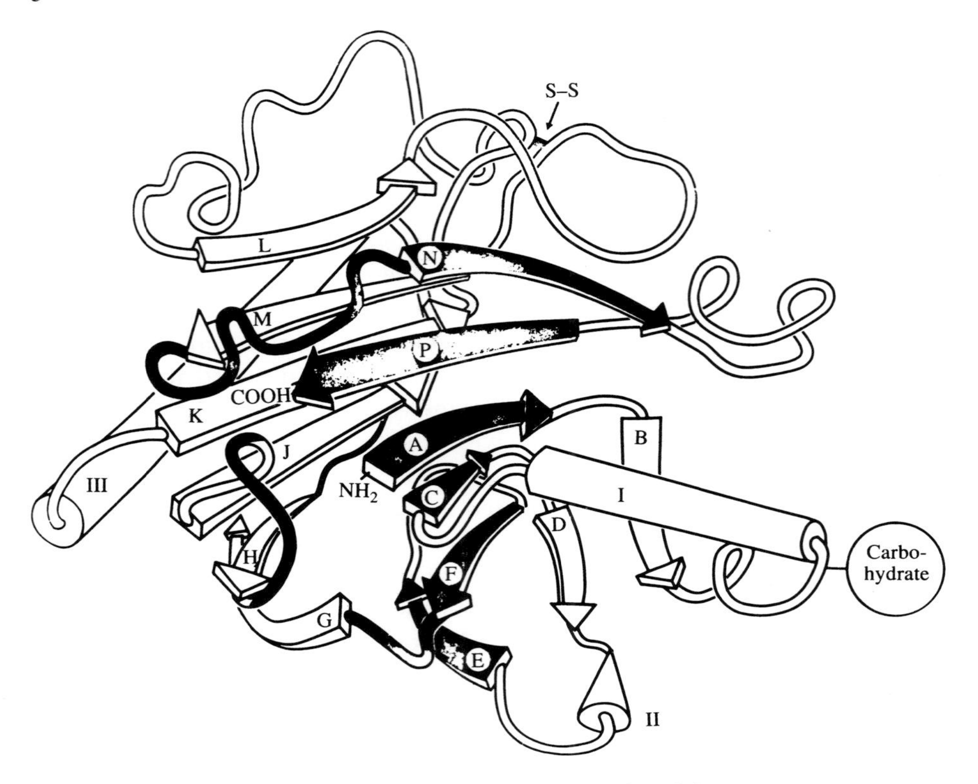

Fig. 37. Schematic drawing of DNase I (from Suck *et al.* 1984).

structure (Drew & Travers, 1984; Klug *et al.* 1979; Suck & Oefner, 1986). The crystal structure of DNase I complexed with a 14 bp DNA duplex (Suck *et al.* 1988) does indeed show the enzyme binding to a distorted DNA that has a minor groove widened by 3 Å and a 21·5° bend of the DNA helix axis away from the protein.

DNase I is a 30400 molecular-weight monomeric endonuclease (Moore, 1981) that requires divalent cations and shows optimal activity in the presence of Ca^{2+} and Mg^{2+} or Ca^{2+} and Mn^{2+} (Price, 1975). The crystal structure of both the enzyme alone and that of a complex with a DNA duplex 14-mer have been refined at 2·0 Å resolution to crystallographic R-factors of 0·16 and 0·18, respectively (Oefner & Suck, 1986; Suck *et al.*, 1988).

DNase I is an α β-protein that contains two 6-stranded primarily anti-parallel β-pleated sheets packed against each other (Suck *et al.* 1984). The four α-helices lie on either side of the β-sheets giving rise to a layered structure that is helix–sheet–sheet–helix (Fig. 37). The two β-sheets and their associated helices each have a very similar structure and are related by a dyad axis between them running along the direction of the strands.

The complex with DNA shows (Suck *et al.* 1988) that the sugar phosphate backbone of the duplex strand which is to be cut lies in a cleft formed between the two β-sheets (Fig. 38). One β-sheet contributes side chains that make numerous interactions with the backbone of the strand to be cleaved and contains His131

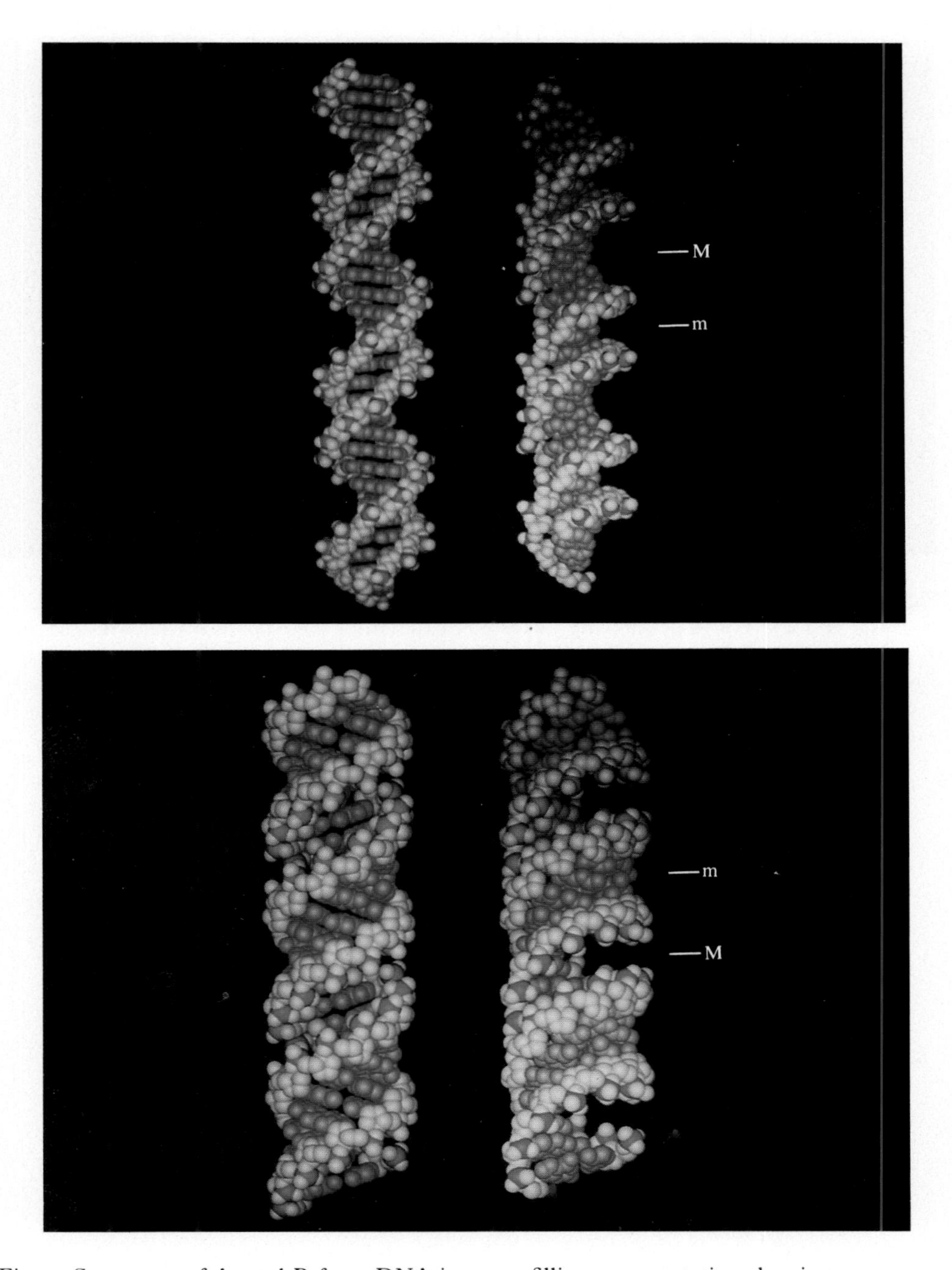

Fig. 1. Structures of A- and B-form DNA in space-filling representation showing differences in major and minor groove widths and shapes. On the top is shown B-DNA and on the bottom A-DNA. In the models on the left the helix axes are parallel to the page whereas on the right the helix axes have been tilted up by 32° to show the groove shapes. The bases are coloured blue, the phosphorous atoms are green and all other atoms are white. The edges of the bases are easily accessible from the major groove of B-DNA and the minor or shallow groove of A-DNA (or RNA). The minor groove is designated by m and the major groove by M.

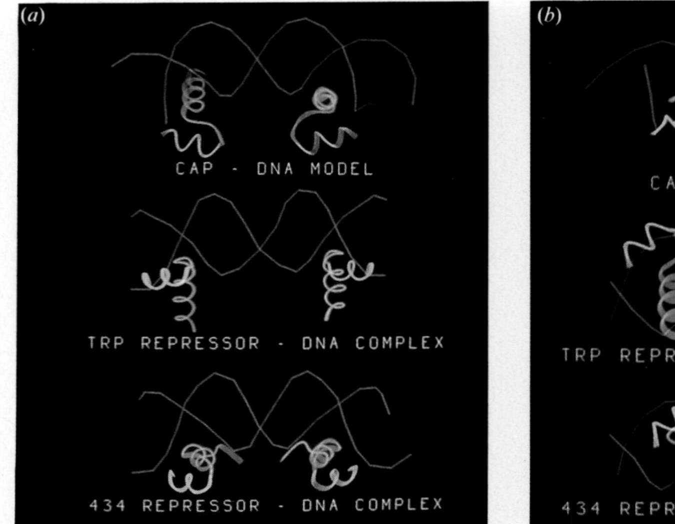

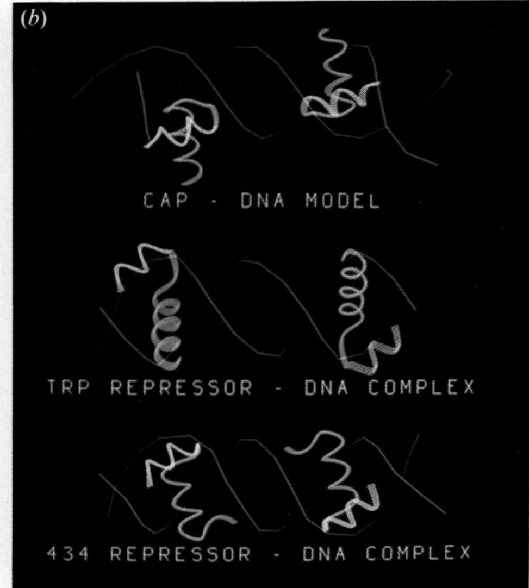

Fig. 4. α-carbon backbone drawing of the DNA-binding domains of CAP, *E. coli* trp repressor and 434 repressor showing that the orientation of the helix–turn–helix relative to the major groove of B-DNA differs significantly in each case. The first helix is yellow and the second, 'recognition' helix, is blue. In part (*a*) the dyad axis is in the plane of the page and in part (*b*) it is perpendicular to the page. The CAP complex is taken from the model-built complex, but the co-crystal structure is very similar at this level of detail.

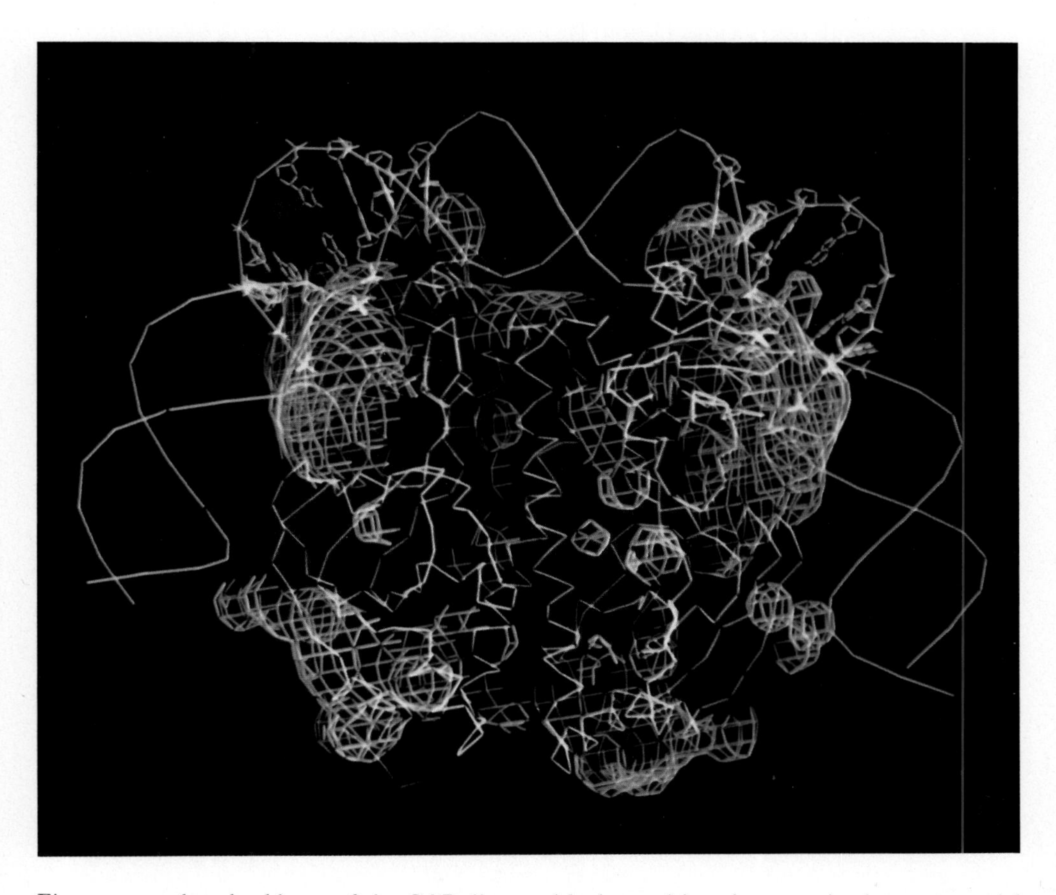

Fig. 11. α-carbon backbone of the CAP dimer with the positive electrostatic charge potential (in blue) contoured at 2 kT. The DNA backbone model (green) built to accommodate the site size and the positive electrostatic potential is shown (from Warwicker *et al.* 1987). Full base pairs surrounding sites at which kinks in the DNA were introduced are shown as well as phosphates (crosses) protected from ethylation by CAP binding.

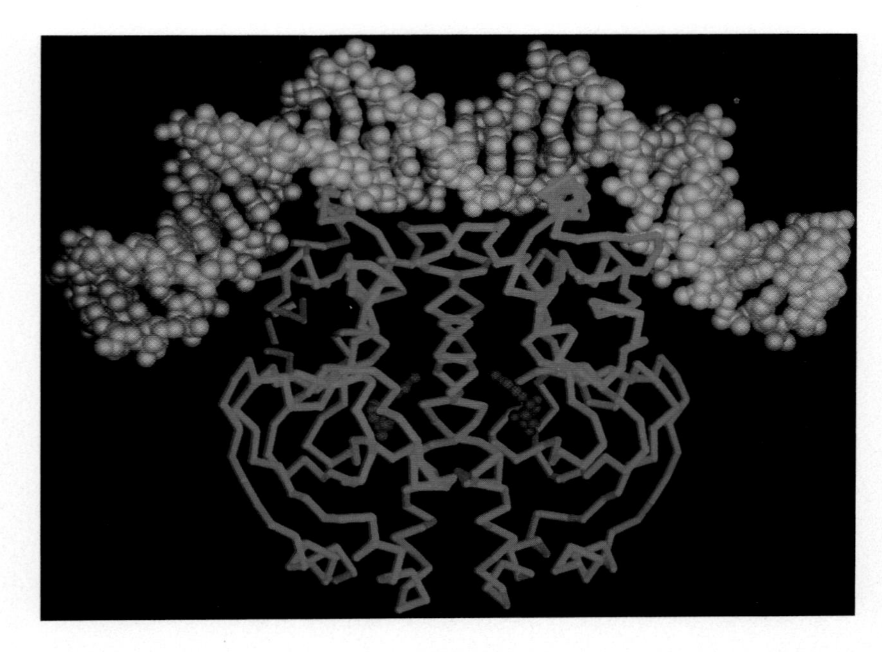

Fig. 12. CAP–DNA complex model derived from co-crystal structure at 3·0 Å resolution (Schultz, Shields & Steitz unpublished). In yellow space-filling representation (using less than van der Waals radii) is the 30 bp duplex DNA and in blue is the α-carbon backbone of the CAP dimer. In red are the two bound cAMP molecules. The DNA is bent by about 90° overall, achieved primarily through two large kinks of about 45° each between base-pairs 5 and 6 from the dyad axis. In the absence of crystal-packing interactions, it is possible that the ends of the DNA will bend towards the protein as model built in Fig. 11. Note that the minor groove at the kink is greatly enlarged. In this view the helix axis of the recognition helix F is perpendicular to the page and parallel to the base-pairs but not the major groove. The protein dimer is held together through the interaction between two long α-helices in a coiled-coil conformation at the dyad axis.

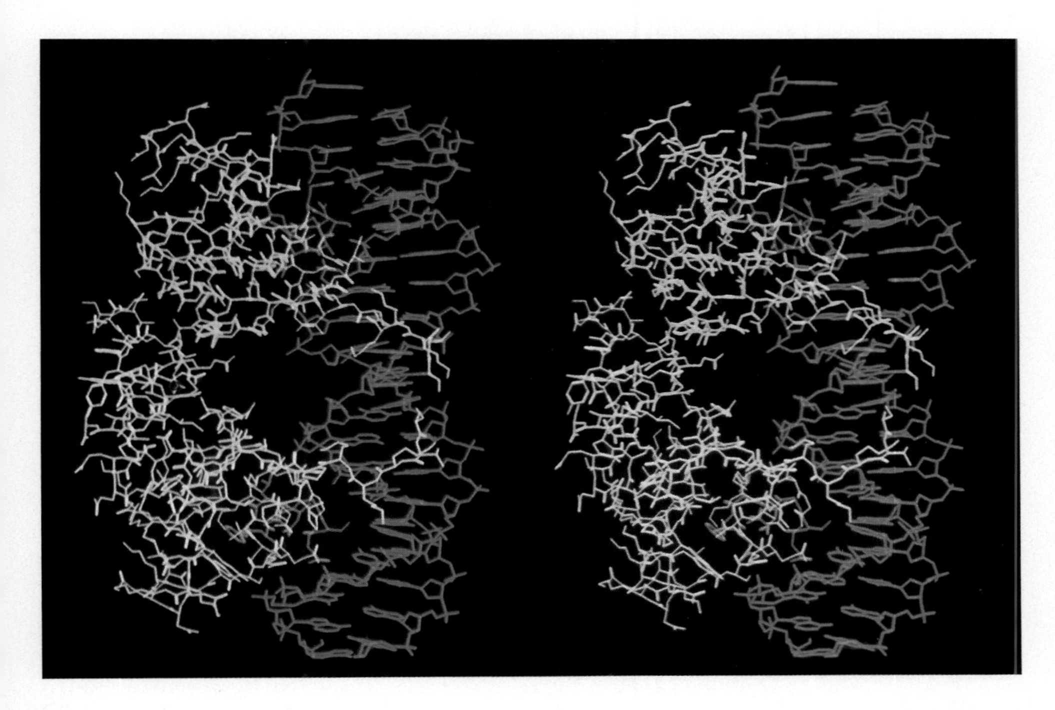

Fig. 23(*a*). Stereo photograph of the λ repressor fragment–operator complex. The DNA is dark blue, the repressor monomer bound to the consensus half-site is yellow and to the non-consensus half-site is purple. The 'recognition' helix (res. 44–52) is red (from Jordan & Pabo, 1988).

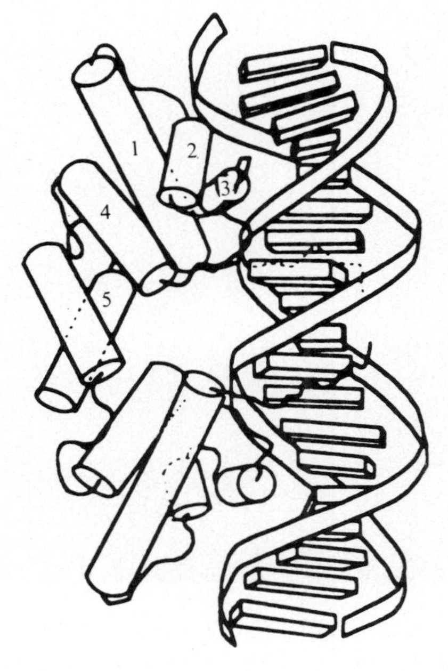

Fig. 23(*b*). Overall structure of λcI repressor fragment dimer interacting with its operator DNA. α-helices are represented as tubes (from Jordan & Pabo, 1988).

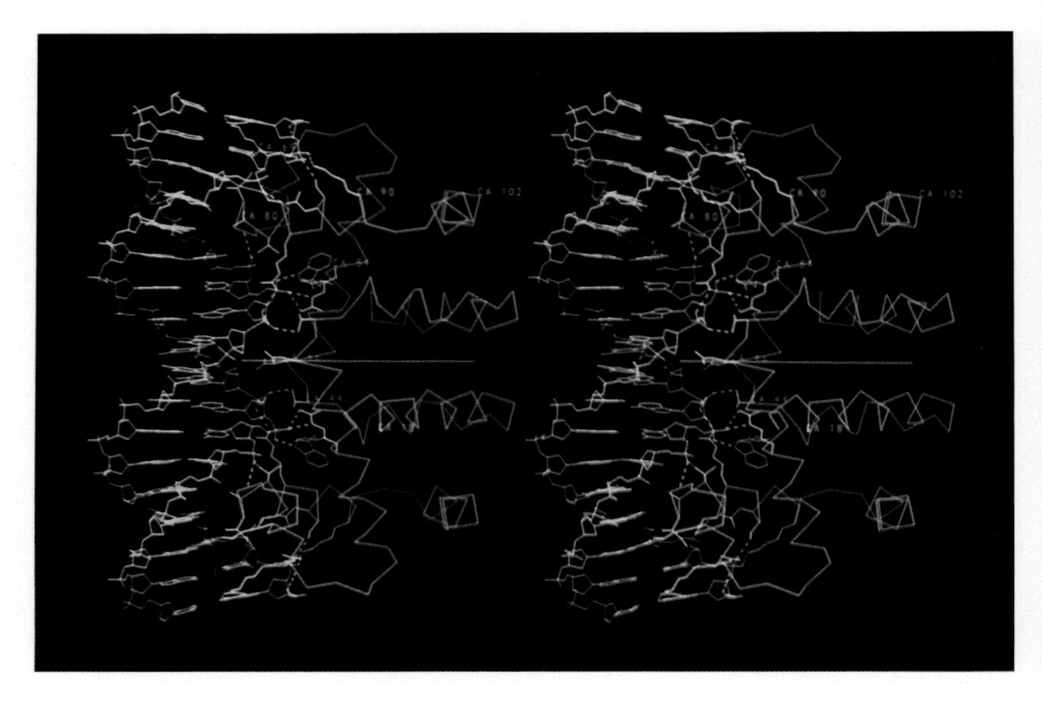

Fig. 28. Colour stereo drawing of TrpR interacting with its operator. Only those protein side-chains seen to be interacting with DNA in the crystal structure are shown (from Otwinowski *et al.* 1988).

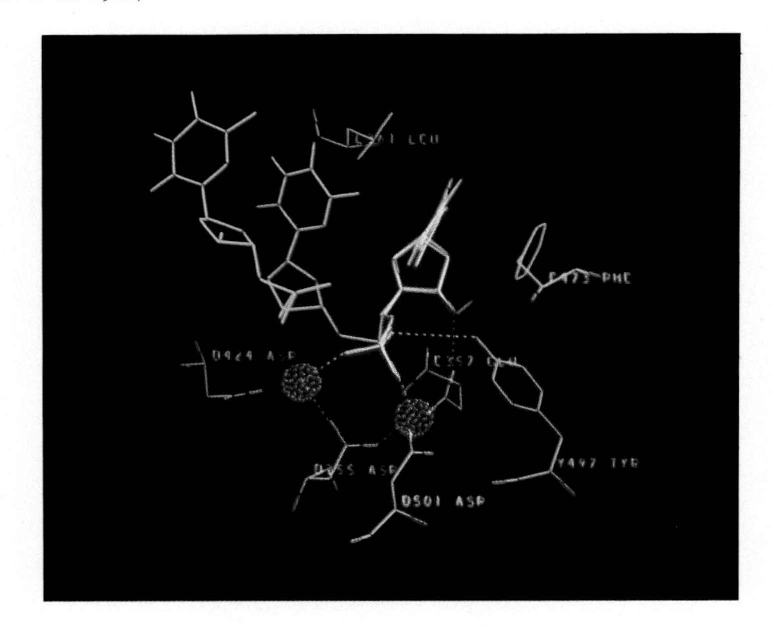

Fig. 36. Model of the deoxytrinucleotide, dT_3 (orange) bound to the exonuclease active site of the Klenow fragment. The 3′ nucleotide is seen to bind nearly identically to the dTMP (superimposed, white). The two metal ions are represented as blue spheres and the H-bonding interactions are shown as dotted lines (Beese and Steitz, unpublished).

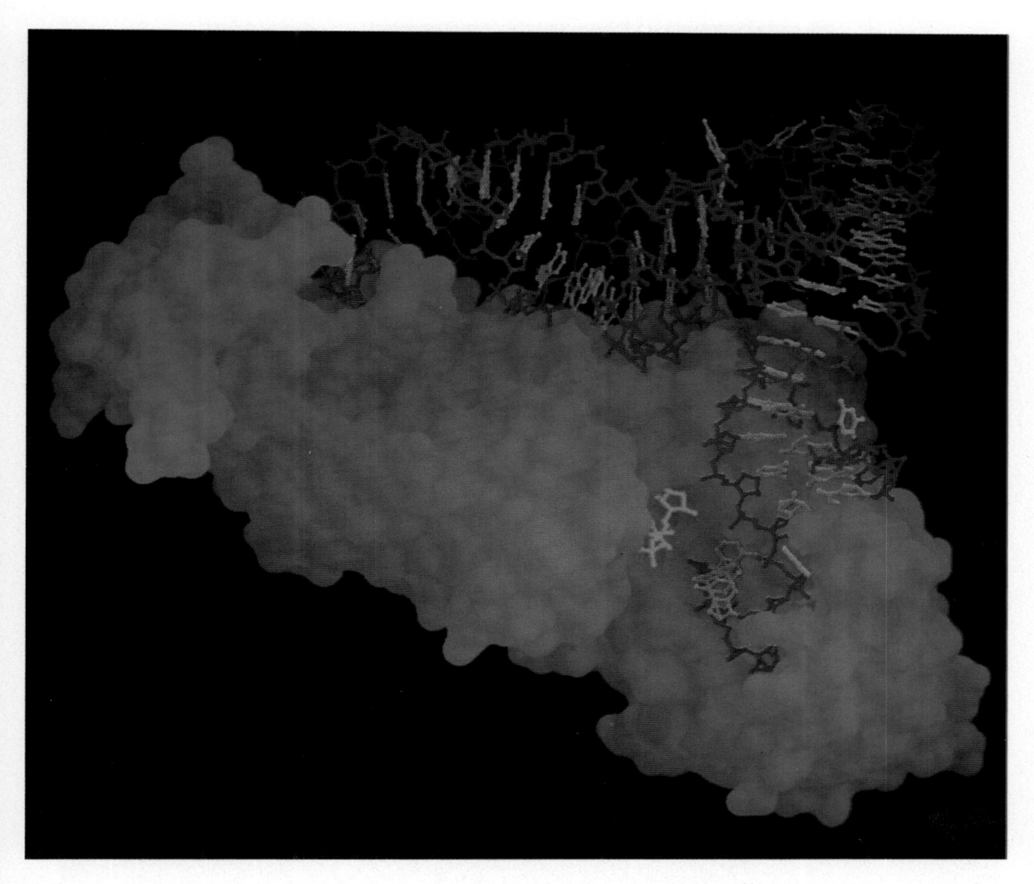

Fig. 41. Solvent accessible surface representation of the GlnRS enzyme complexed with $tRNA^{Gln}$ and ATP. The region of contact between tRNA and protein extends across one side of the entire enzyme surface and includes interactions from all four protein domains. The acceptor end of the tRNA and the ATP are seen in the bottom of a deep cleft. Protein is inserted between the 5′ and 3′ ends of the tRNA and disrupts the expected base pair between U1 and A72 (from Rould *et al.* 1989).

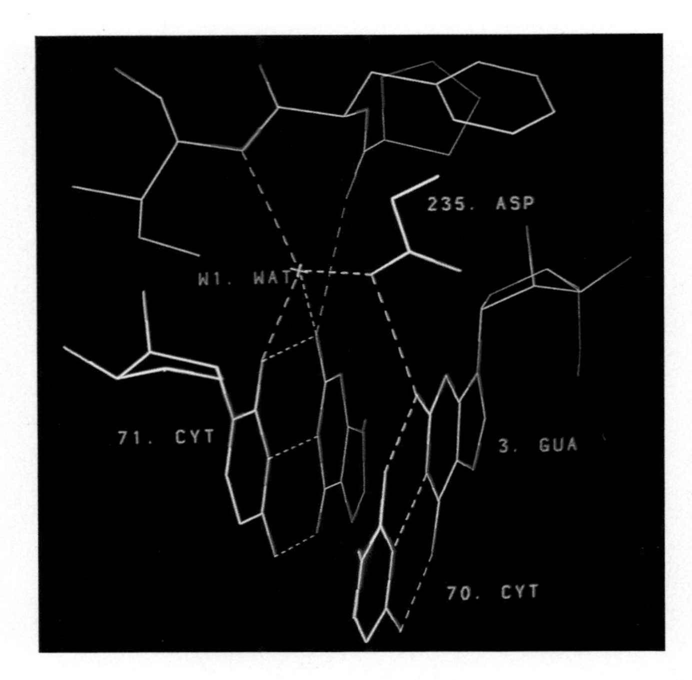

Fig. 44. Hydrogen bonding interactions between GlnRS and the edges of base pairs G2–C71 and G3–C70 exposed in the minor groove. The complementary H-bonding surface includes a buried water molecule (W1. WAT). (Rould, Perona & Steitz, unpublished).

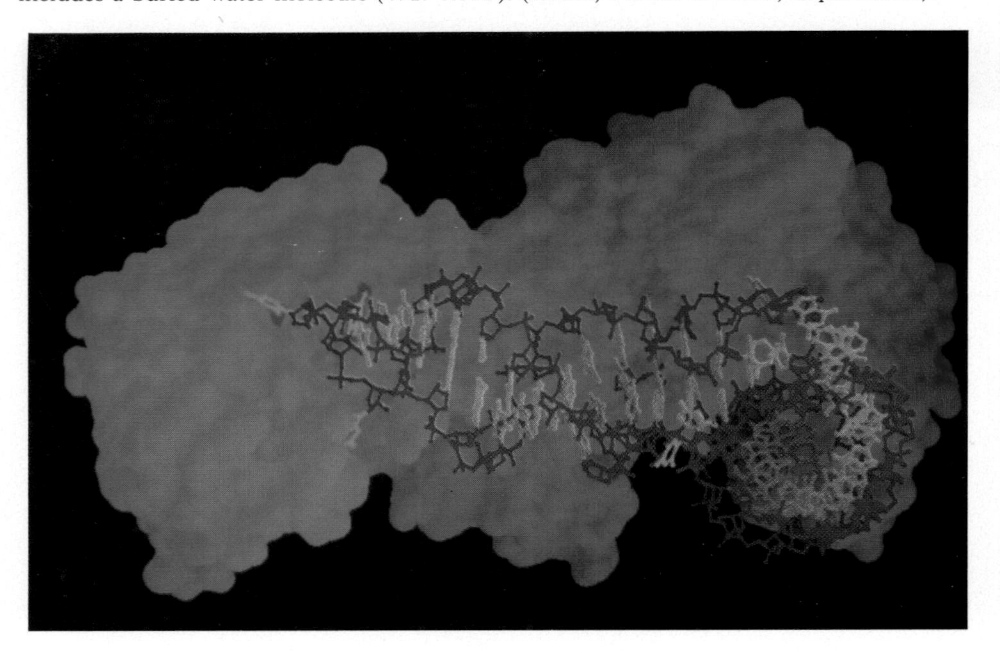

Fig. 45. Solvent accessible surface representation of the GlnRS enzyme complexed with $tRNA^{Gln}$ and ATP as viewed 'down' the acceptor stem. The three bases of the anticodon are seen to be splayed out and interacting with the protein. The anticodon bases are coloured green. (Rould, Perona & Steitz, unpublished).

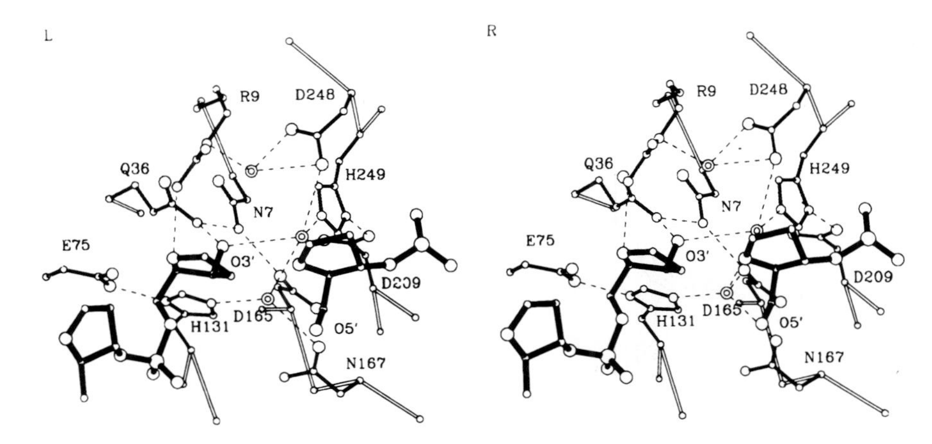

Fig. 38. Stereo drawing of the protein environment of a DNA cutting site in the DNase I co-crystal with DNA at the nick in the 14-mer duplex. The essential His131 is located in the neighbourhood of the expected position of the phosphate. Double circles represent water molecules and heavy lines indicate the DNA backbone (from Suck *et al.* 1988).

thought to function as a general base in catalysis (Suck & Oefner, 1986). The other β-sheet protrudes into the minor groove, binds the essential Ca^{2+} and makes backbone interactions with the non-cleaved strand. A β-loop from this sheet puts Tyr73 and Arg38 into the narrow groove of the B-form DNA which requires a widened narrow groove.

No conformational change occurs in Dnase I upon binding of the oligonucleotide, except for the positions of a few side chains that contact the DNA directly. The C^{α}-positions of the complexed and uncomplexed enzyme superimpose with an RMS difference of 0·37 Å.

In contrast to the relatively invariant structure of the protein upon complexation, the DNA is significantly distorted from standard B-type DNA (Suck *et al.* 1988). A major distortion is induced by a stacking interaction between the side chain of Tyr73 and the deoxyribose ring of the nucleotide on the 5′ side and penultimate to the cleavage site. Changes in the deoxyribose affects the base orientation which in turn results in a high individual roll angle (13·1°) between that base and the next one at the cleavage site. There is a 21·5° bend towards the major groove (away from the enzyme). The minor groove is widened to about 15 Å from the canonical B-DNA.

The requirement of the enzyme for a wider narrow groove adjacent to the cleavage site would account for the lower cutting rates on AT-rich sequences since these sequences tend to have narrower than average minor grooves (Yoon *et al.* 1988).

6.3 E. coli *Hu protein*

The *E. coli* Hu protein is a 9500 molecular-weight protein that forms a dimer that binds to duplex DNA in a non-sequence-specific fashion inducing the formation

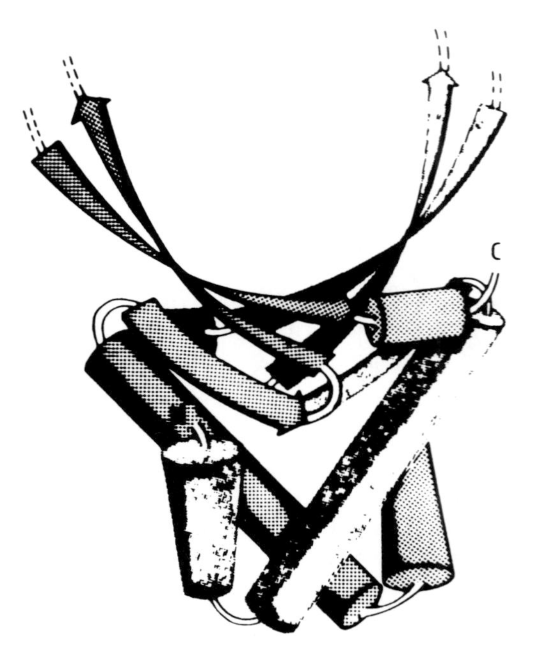

Fig. 39. A schematic representation of the Hu dimer with each subunit shaded differently (from Tanaka *et al.* 1984).

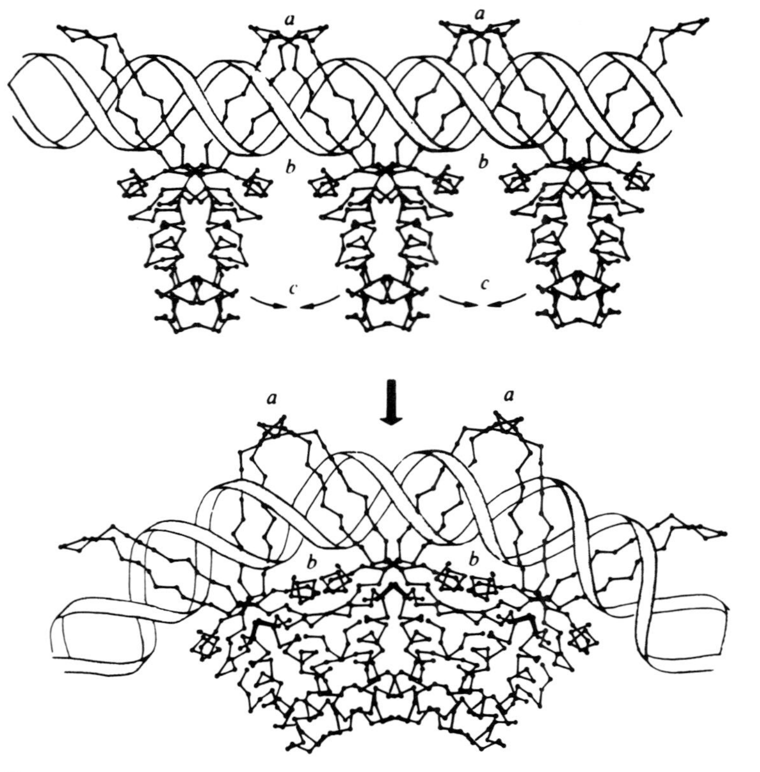

Fig. 40. A proposal by Tanaka *et al.* (1984) for how Hu might interact with B-DNA using its two anti-parallel β-ribbon 'arms' and how this protein might induce supercoiling of the DNA (from Tanaka *et al.* 1984).

of bead-like structures in a manner analogous to the histones (Rouvière-Yaniv & Yaniv, 1979). Its function appears to be in condensing the *E. coli* chromosome. The crystal structure of this protein reveals a dimer that is held together by α-helices (Tanaka *et al.* 1984). An anti-parallel β-loop from each subunit from outstretched arms that create a large cleft sufficient in size to accommodate duplex DNA (Fig. 39). The tips of these anti-parallel β-loops are disordered in the crystal structure. While this duplex DNA-binding protein has a helix–turn–helix structure quite similar to that observed in the gene regulatory proteins, it appears, at least in the crystal structure, to be involved in the formation of dimer and thus not available for interaction with DNA.

A model for an Hu–DNA complex has been presented (Tanaka *et al.* 1984) in which the anti-parallel β-loops of the two subunits encompass the DNA either in the major groove or in the minor groove (Fig. 40) in a manner proposed earlier for anti-parallel β-strands on the basis of model building studies (Carter & Kraut, 1974; Church *et al.* 1977). A model has also been presented for condensation of the DNA by this protein noting that a constant curvature is introduced by dimers placed side by side.

A protein showing strong amino acid sequence homology, integration host factor (IHF), binds to DNA in a sequence specific fashion. Thus, a protein having a very similar structure to that of Hu is able to recognize DNA presumably using anti-parallel β-strands if the model for Hu–DNA interaction is confirmed. A proposal that IHF makes sequence specific contacts in the minor groove via these anti-parallel β-strands has been made by Yang & Nash (1989) to account for methylation protection studies of IHF binding to DNA.

7. SEQUENCE-SPECIFIC RNA-BINDING PROTEINS

7.1 E. coli *glutaminyl-tRNA synthetase complexed with* $tRNA^{Gln}$

An interesting problem in RNA sequence discrimination is presented by the 20-aminoacyl tRNA synthetases that translate the genetic code by correctly charging a specific amino acid onto a tRNA containing the appropriate anti-codon. Since the 40 or more different tRNAs in a cell have the same overall 3D structure and must function interchangeably in protein synthesis on the ribosome, the ability of these enzymes to discriminate accurately among these similar RNA molecules provides an interesting challenge in understanding the structural basis of sequence recognition.

The structures of two aminoacyl tRNA synthetases, those of *B. stearothermophylus* tyrosyl-tRNA synthetase (TyrRS) (Bhat *et al.* 1982) and *E. coli* methional-tRNA synthetase (MetRS) (Zelwer *et al.* 1982) have been determined as complexes with either ATP or amino-acid substrates. Comparison of the structures of these two enzymes revealed a similar 5-stranded parallel β-sheet of a 'dinucleotide-binding' motif that had previously been observed in virtually all dehydrogenase and kinase structures (Rossmann *et al.* 1975). The crystal structures of $tRNA^{Phe}$, $tRNA^{Asp}$ and $tRNA^{fMet}$ revealed that they all share a

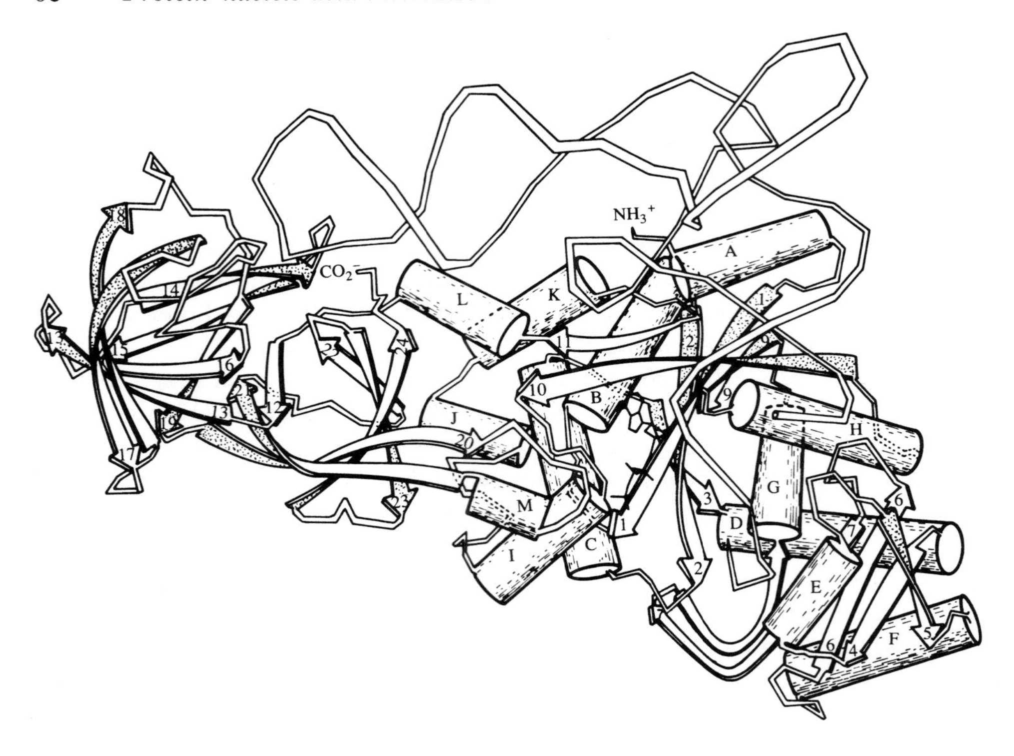

Fig. 42. Schematic backbone drawing of GlnRS complexed with tRNA^{Gln} and ATP. For the protein, α-helices are represented as tubes sequentially lettered and β-strands as arrows sequentially numbered. The dinucleotide fold domain includes residues from the β-strand 1–3 plus α-helix G to β-strand 10. The acceptor end-binding domain includes chain between the amino end of α-helix D to the carboxyl end of β-strand 8. The two β-barrel anticodon-binding domains consist of β-strand B to β-strand 19 for the distal domain and β-strand 20 to the carboxyl terminus for the proximal domain.

common and now well known L-shaped structure (Robertus *et al.* 1974; Kim *et al.* 1974; Moras *et al.* 1980; Woo *et al.* 1980). While some differences exist in the conformation of the acceptor end and in the overall bend between the two halves, these structures seem very similar indeed and suggest that all tRNAs may have approximately the same structures and solution.

The *E. coli* glutaminyl-tRNA synthetase (GlnRS) is a 64000 molecular-weight monomer of 553 amino acids. The cloning and overexpression of both the GlnRS and tRNA$_2^{Gln}$ allowed purification of large quantities of both of these macromolecules and their co-crystallization with ATP and magnesium (Perona *et al.* 1989). The crystal structure of this monomeric complex has been solved at 2·8 Å resolution and partially refined at 2·5 Å (Rould *et al.* 1989). GlnRS is observed to interact extensively with tRNA^{Gln} from the acceptor end to the anticodon, including the inside of the L (Fig. 41).

The monomeric protein consists of four domains arranged to give an elongated molecule with an axial ratio of greater than 3 to 1 (Fig. 42). The most amino terminal of these domains forms a 5-stranded parallel β-sheet folded in the manner known as the dinucleotide (or Rossmann) fold. This domain is seen to bind ATP, glutamine and the 3′ end of the tRNA. Furthermore, this domain makes at least one sequence-specific interaction with base pairs 3–70. Because the dinucleotide fold domain has all the rudimentary elements of a synthetase, it might therefore correspond to the primordial synthetase domain common to all 20 synthetases. The overall structure of the dinucleotide fold domain is very similar in GlnRS to that found in both the TyrRS and the MetRS enzymes as is its interaction with the amino acid and ATP substrates.

A second domain is inserted between the two halves of the dinucleotide fold domain. This domain, which consists of both β-sheet and α-helix is responsible for many interactions in the acceptor stem of the tRNA. The two domains that are nearest to the C-terminus are β-barrel structures. These β-barrel motifs are interacting with the anticodon stem and loop.

(i) *tRNA^{Gln} structure*. A comparison of the phosphate backbone structure of tRNA^{Gln} in this complex with that of the uncomplexed yeast tRNA^{Phe} shows several major differences, the most dramatic of which is in the acceptor stem and the 3′ end. The most striking difference in the tRNA^{Gln} is that a base pair between nucleotide 1 and 72 is disrupted and the 3′ CCA end of the tRNA hairpins back in the direction of the anticodon rather than continuing on from the acceptor stem in the helical fashion (Fig. 43). The bases of A76, C75 and G73 are stacked on each other and the nucleotides C74 is looped out with its base interacting with a complementary pocket in the protein. The hairpinned structure of the 3′ end is stabilized in part by a hydrogen bond between the 2-amino group of G73 and the phosphate backbone of residue 72 of the tRNA and in part by numerous interactions with the protein. A second major difference between these two tRNA molecules involves the conformation of the anticodon loop which has opened out in the complex allowing the bases of the anticodon to interact extensively and directly with the protein.

(ii) *tRNA discrimination*. Molecular-genetic studies, both *in vivo* and *in vitro*, on the specificity of charging of tRNAs have shown that a relatively small number of nucleotides can be important for the ability of synthetases to select their cognate tRNA correctly (Yarus, 1988; Schulman & Abelson, 1988; Normanly & Abelson, 1989). Both by the study of mischarging of mutant suppressor tRNAs *in vivo* and the charging of T7 tRNA transcripts *in vitro* some of the identity elements in tRNAs have been identified. In the case of the GlnRS the major identity determinants that have emerged from molecular-genetic studies thus far are U35 in the anticodon (Yaniv *et al.* 1974; Schulman & Pelka, 1985) and G73 (Shimura *et al.* 1972; Hooper *et al.* 1972; Seong *et al.* 1989) and base-pair 1–72 (Hooper *et al.* 1972; Seong *et al.* 1989) in the acceptor stem. The crystal structure of the complex strongly supports these as recognition elements and additionally suggests an important role in tRNA discrimination for base pairs 3–70 and 2–71 in the acceptor stem and the bases 34, 35 and 36 of the anticodon.

The structural explanation for the apparent preference of GlnRS for a G at position 73 is that only guanine can make the observed hydrogen bond between its 2-amino group and the phosphate oxygen of the previous nucleotide (Fig. 43).

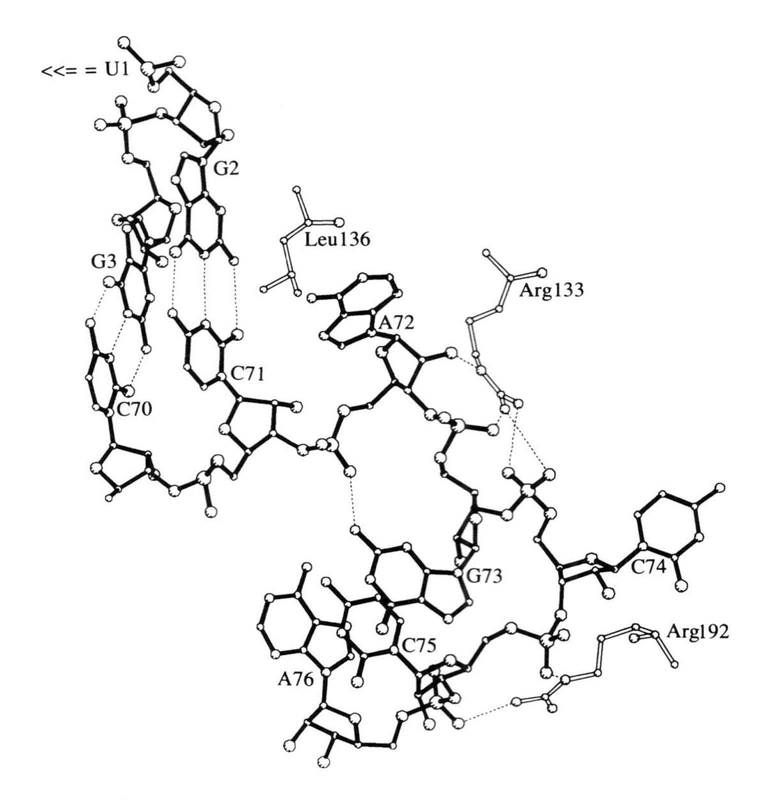

Fig. 43. The acceptor strand of the tRNA as seen in the complex. The side chain of Leu136 extends from a β turn and wedges between the bases of nucleotides A72 and G2, disrupting the last base pair of the acceptor stem, U1-A72. The enzyme stabilizes the hairpin conformation via the interaction of several basic side chains with the sugar-phosphate backbone. An intramolecular hydrogen bond between the 2-amino group of G73 and the phosphate group of A72 further stabilizes this conformation. (From Rould *et al.* 1989.)

This hydrogen-bonding interaction serves to stabilize the hairpinned 3′ end of the tRNA, a structure that is complementary to the enzyme-binding site. This base appears to be an important recognition element by virtue of an RNA–RNA interaction rather than an RNA–protein interaction.

A structural basis for the observation that AU or mismatched base pairs are more favoured by GlnRS at position 1–72 than GC seems clear. The free energy cost of breaking either a mismatched base pair of an AU base pair is less than for breaking a GC base pair. As in the case of G73, specificity of recognition is achieved because the free energy cost for the tRNA to take up the conformation required for complementarity to the enzyme depends on the sequence of the RNA. In this case there are not sequence-specific interactions between the bases and the protein.

Two base pairs in the acceptor stem are directly recognized by virtue of protein interaction in the minor or shallow groove of the tRNA (Fig. 44). The surface of the protein that is complementary to these two base pairs in the minor groove is

Arg Gln Glu Uracil 35 Cytosine

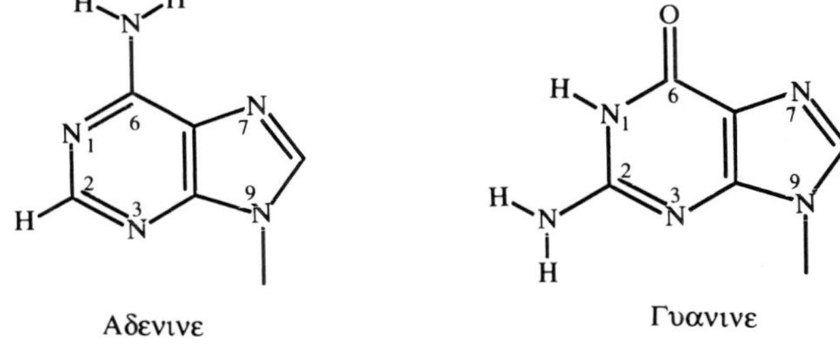

Αδενινε Γυανινε

Fig. 46. Schematic drawing of the hydrogen-bonding interactions between U_{35} of $tRNA^{Gln}$ and GlnRS (M. Rould, J. Perona, T. Steitz, unpublished). The protein-binding pocket specifies uracil rather than cytosine by virtue of the ionic hydrogen bonds – Arg to O4 and Glu to N3. Adenine and guanine would not fit into the binding site.

presented by the end of an anti-parallel β-loop and an aspartic acid emanating from the amino end of α-helix H (Fig. 44). A hydrogen bond between the backbone carbonyl of Pro181 and the N2 of guanine 2 is seen as well as a hydrogen bond between the β-carboxylate group of Asp235 and the N2 of guanine 3. In addition to these direct interactions between the protein and the minor groove there is a water-mediated set of hydrogen bonds with these two base pairs and the backbone amide of residue 183 and the carboxylate of Asp235. This water molecule is an obligate H-bond acceptor from the N2 of G2 because of its other H-bond interactions.

(iii) *Anticodon recognition*. There is extensive interaction between the three bases of the anticodon of $tRNA^{Gln}$ and the protein (Fig. 45). The three anticodon bases are unstacked and splayed out – each binding to a separate protein 'pocket' (Rould, Perona & Steitz, unpublished). The second two bases of the anticodon – U_{35} and G_{36} – are hydrogen bonded to the protein in a manner that is specific to those bases only (Fig. 46). The hydrogen-bonding interactions with C_{34} presumably can be charged to accommodate a U_{34} (as required to a charge $tRNA_{1}^{Gln}$) but not a G or A. Not only are the hydrogen-bonding patterns with the anticodon bases specific for those bases, but they are likely to be high-energy hydrogen bonds. In all cases buried, charged-protein side chains are involved – Arg and Glu with U_{35}, and Arg with G_{36} and 2 Arg with C_{34}.

8. CONCLUSIONS AND PROSPECTS

It appears that proteins are able to directly specify the duplex DNA sequence to which they specifically bind by protruding an α-helix (or two anti-parallel β-strands) into the major groove of B-form DNA. The side chains emanating from the secondary structure elements, including water molecules that become buried upon formation of the complex, are complementary to the shape and hydrogen-bonding pattern that the correct DNA sequence presents to the major groove and not complementary to that of the incorrect DNA sequence. In the case of protein recognition of duplex RNA sequence, it appears that the protein presents a surface complementary in structure to features that the exposed edges of base pairs present to the minor groove of the A-form RNA. Additionally, a protein is able to directly recognize bases in single stranded regions of the RNA by forming a pocket complementary in detail to a specific base.

Nucleic-acid sequences can also be recognized by virtue of the different distortability of different nucleic-acid sequences. Overall it appears that upon formation of a complex with DNA the change in protein structure that occurs are modest (except for the changes in interdomain relationships that occur when CAP and *cro* bind to DNA). The changes in the nucleic-acid structure upon complex formation, however, appear in many cases to be dramatic as exemplified in the DNA complexes with *Eco*R I, CAP, 434 repressor, DNase I and the RNA complex between GlnRS and $tRNA^{Gln}$. The protein structures provide a rack to which the nucleic-acid structures must conform in order to bind.

(i) *Future problems*. While the principles by which proteins are able to discriminate among various nucleic-acid sequences are becoming reasonably well elucidated, an understanding of how the wide panorama of nucleic-acid-binding proteins achieve their biological functions subsequent to binding is less well advanced. Many nucleic-acid-binding proteins work as a part of a larger assembly of proteins. A particularly striking example of the complexity of multiprotein assemblies is provided by the question of how transcription activator proteins are able to activate RNA polymerase. In the simplest cases proteins such as *E. coli* CAP bound to a single site can activate RNA polymerase. In the more complex eukaryotic transcription systems the binding of multiple copies of each of several different transcription factors work in concert to activate RNA polymerase. The only way to provide a structural basis for understanding these processes is to know the 3D structure of both the activating proteins and the polymerase bound to the relevant piece of duplex DNA. Other large assemblies of proteins are involved in DNA replication, recombination, topoisomerase and gyrase activity, RNA splicing and the ribosome. Presently the only structural technologies that would allow detailed determination of the structures of such large-molecular-weight assemblies are X-ray crystallography and high-resolution electron microscopy of 2D crystals.

A second problem for future study involves understanding the structural basis of energy driven relative motion of macromolecules. Examples from both general and site-specific recombination require the rearrangement of duplex DNAs bound to the recombination proteins. A final striking example is provided by proteins that catalyse topological changes in the DNA, the topoisomerases and gyrases. Associated with these problems are the reasons for and the nature of DNA wrapping around the protein assemblies.

9. ACKNOWLEDGEMENTS

I wish to thank C. Pabo, S. Harrison, S. Phillips, D. Suck, B. Matthews, M. Ptashne, P. Sigler, J. Berg and J. Rosenberg for providing manuscripts before their publication and for useful discussions. I thank Paul Vogt for assistance in preparing several of the coloured figures.

10. REFERENCES

ABDEL-MEGUID, S. S., GRINDLEY, N. D. F., SMYTH TEMPLETON, N. & STEITZ, T. A. (1984). Cleavage of the site-specific recombination protein $\gamma\delta$ resolvase: the smaller of two fragments binds DNA specifically. *Proc. natn. Acad. Sci. USA* **81**, 2001–2005.

ABDEL-MEGUID, S. S., MURTHY, H. M. K. & STEITZ, T. A. (1986). Preliminary X-ray diffraction studies of the putative catalytic domain of $\gamma\delta$ resolvase from *Escherichia coli*. *J. biol. Chem.* **261**, 15934–15935.

ADLER, K., BEYREUTHER, K., FANNING, E., GEISLER, N., GRONENBORN, B., KLEMM, A., MÜLLER-HILL, B., PFAHL, M., SCHMITZ, A. (1972). How *Lac* repressor binds to DNA. *Nature* **237**, 322–326.

AGGARWAL, A. K., RODGERS, D. W., DROTTAR, M., PTASHNE, M. & HARRISON, S. C. (1988). Recognition of a DNA operator by the repressor of phage 434: a view at high resolution. *Science* **242**, 99–07.

AIBA, H. (1983). Autoregulation of the *Escherichia coli crp* gene: CRP is a transcriptional repressor for its own gene. *Cell* **32**, 141–149.

AIBA, H., FUJIMOTO, S. & OZAKI, N. (1982). Molecular cloning and nucleotide sequencing of the gene for *E. coli* cAMP receptor protein. *Nucl. Acids Res.* **10**, 1345.

AIBA, H., NAKAMURA, T., MITANI, H. & MORI, H. (1985). Mutations that alter the allosteric nature of cAMP receptor protein of *Escherichia coli*. *EMBO J.* **4**, 3329–3332.

ANDERSON, J., PTASHNE, M. & HARRISON, S. C. (1984). Co-crystals of the DNA-binding domain of phage 434 repressor and a synthetic phage 434 operator. *Proc. natn. Acad. Sci. USA* **81**, 1307–1311.

ANDERSON, J. E., PTASHNE, M. & HARRISON, S. C. (1985). A phage repressor-operator complex at 7 Å resolution. *Nature* **316**, 596–601.

ANDERSON, J. E., PTASHNE, M. & HARRISON, S. C. (1987). Structure of the repressor–operator complex of bacteriophage 434. *Nature* **326**, 846–852.

ANDERSON, W. F., CYGLER, M., VANDONSELAAR, M., OHLENDORF, D. H., MATTHEWS, B. W., KIM, J. & TAKEDA, Y. (1983). Crystallographic data for complexes of the *cro* repressor with DNA. *J. molec. Biol.* **168**, 903–906.

ANDERSON, W. F., OHLENDORF, D. H., TAKEDA, Y. & MATTHEWS, B. W. (1981). Structure of the *cro* repressor from bacteriophage λ and its interaction with DNA. *Nature* **290**, 754–758.

ANDERSON, W. F., TAKEDA, Y., OHLENDORF, D. H. & MATTHEWS, B. W. (1982). Proposed α-helical super-secondary structure associated with protein-DNA recognition. *J. molec. Biol.* **159**, 745–751.

BENSON, N., SUGIONO, P. & YONDERIAN, P. (1988). DNA sequence determinants of λ repressor binding *in vivo*. *Genetics* **118**, 21–29.

BERG, J. M. (1988). Proposed structure for the zinc-binding domains from transcription factor IIIA and related proteins. *Proc. natn. Acad. Sci. USA* **85**, 99–102.

BESSE, M., VON WILCKEN-BERGMANN, B. & MÜLLER-HILL, B. (1986). Synthetic lac operator mediates repression through *lac* repressor when introduced upstream and downstram from *lac* repressor. *EMBO J.* **5**, 1377–1381.

BHAT, T. N., BLOW, D. M. & BRICK, P. (1982). Tyrosyl-tRNA synthetase forms a mononucleotide-binding fold. *J. molec. Biol.* **158**, 699–709.

BOELENS, R., SCHEEK, R. M., VAN BOOM, J. H. & KAPTEIN, R. (1987). Complex of *lac* repressor headpiece with a 14 base-pair *lac* operator fragment studied by two-dimensional nuclear magnetic resonance. *J. molec. Biol.* **193**, 213–216.

BOGENHAGEN, D. F., SAKONJU, S. & BROWN, D. D. (1980). A control region in the centre of the 5S RNA gene directs specific initiation of transcription: the 3′ border of the region. *Cell* **19**, 27–35.

BRAYER, G. D. & MCPHERSON, A. (1983). Refined structure of the gene 5 DNA binding protein from bacteriophage fd. *J. molec. Biol.* **169**, 565–596.

BRENNAN, R. G., TAKEDA, Y., KIM, J., ANDERSON, W. F. & MATTHEWS, B. W. (1986). Crystallization of a complex of *cro* repressor with a 17 base-pair operator. *J. molec. Biol.* **188**, 115–118.

BRICK, P., BHAT, T. N. & BLOW, D. M. (1989). Structure of tyrosyl-tRNA synthetase refined at 2·3 Å resolution. Interaction of the enzyme with the tyrosyl adenylate intermediate. *J. molec. Biol, in press.*

BRICOGNE, G. (1976). Methods and programs for direct-space exploitation of geometric redundancies. *Acta Crystallogr.* **A32**, 832.

BRUNIE, S., MELLOT, P., ZELWER, C., RISLER, J.-L., BLANQUET, S. & FAYAT, G. (1987). Structure-activity relationships of methionyl-tRNA synthetase: graphics modelling and genetic engineering. *J. molec. Graphics* **5**, 18–28.

BRUTLAG, D., ATKINSON, M. R., SETLOW, P. & KORNBERG, A. (1969). An active fragment of DNA polymerase produced by proteolytic cleavage. *Biochem. biophys. Res. Comm.* **37**, 982–989.

BRUTLAG, D. & KORNBERG, A. (1972). Enzymatic synthesis of doexyribonucleic acid. *J. biol. Chem.* **247**, 241–248.

BURLINGAME, R. W., LOVE, W. E., WANG, B.-C., HAMLIN, R., XUONG, N. H. & MOUDRIANANKIS, E. N. (1985). Crystallographic structure of the octameric histone core of the nucleosome at a resolution of 3·3 Å. *Science* **228**, 546–553.

Carter, C. W. & Kraut, J. (1974). A proposed model for interaction of polypeptides with RNA. *Proc. natn. Acad. Sci. USA* **71**, 283–287.

Charlier, B. M., Maurizot, J. C. & Zaccui, G. (1980). Neutron scattering studies of *lac* repressor. *Nature* (*London*) **286**, 423–425.

Church, G. M., Sussman, J. L. & Kim, S.-H. (1977). Secondary structure complementarity between DNA and proteins. *Proc. natn. Acad. Sci. USA* **74**, 1458–1462.

Coll, M., Frederick, C. A., Wang, A. H.-J. & Rich, A. (1987). A bifurcated hydrogen-bonded conformation in the d(AT) base pairs of the DNA dodecamer d(CGCAAATTTGCG) and its complex with distamycin. *Proc. natn. Acad. Sci. USA* **84**, 8385–8389.

Cossart, P. & Gicquel-Sanzey, B. (1982). Cloning and sequence of the crp gene of *Escherichia coli* K12. *Nucl. Acids Res.* **10**, 1363–1378.

La Cour, T. F. M., Nyborg, J., Thirup, S. & Clark, B. F. C. (1985). Structural details of the binding of guanosine diphosphate to elongation factor Tu from *E. coli* as studies by X-ray crystallography. *EMBO J.* **4**, 2385–2388.

Crick, F. H. C. & Klug, A. (1975). Kinky helix. *Nature* **255**, 530–533.

De Crombrugghe, B., Busby, S. & Buc, H. (1984). Cyclic AMP receptor protein: role in transcription activation. *Science* **224**, 831–838.

Delarue, M. & Moras, D. (1989). RNA structure. In *Nucleic Acids and Molecular biology*, vol. 3 (ed. F. Eckstein and D. M. J. Lilley), pp. 182–196. Springer-Verlag.

Derbyshire, V., Freemont, P. S., Sanderson, M. R., Beese, L. S., Friedman, J. M., Steitz, T. A. & Joyce, C. M. (1988). Genetic and crystallographic studies of the 3′,5′-exonucleolytic site of DNA polymerase I. *Science* **240**, 199–201.

Diakun, G. P., Fairall, L. & Klug, A. (1986). EXAFS study of the zinc-binding sites in the protein transcription factor IIIA. *Nature* **324**, 698–699.

Dickerson, R. E. (1983). Base sequence and helix structure variation in B- and A-DNA. *J. molec. Biol.* **166**, 419–441.

Dickerson, R. E. & Drew, H. R. (1981). Structure of a B-DNA dodecamer. II. Influence of base sequence on helix structure. *J. molec. Biol.* **149**, 761–786.

Dickson, R. C., Abelson, J., Barnes, W. M. & Reznikoff, W. S. (1975). Genetic regulation: the *lac* control region. *Science* **187**, 27–35.

Digabriele, A. D., Sanderson, M. R. & Steitz, T. A. (1989). Crystal lattice packing is important in determining the bend of a DNA dodecamer containing an adenine tract. *Proc. natn. Acad. Sci. USA* **85**, 1816–1820.

Drew, H. R. & Travers, A. A. (1984). DNA structural variations in the *E. coli tyr*T promoter. *Cell* **37**, 491–502.

Ebright, R. H., Cossart, P., Gicquel-Sanzey, B. & Beckwith, J. (1984). Mutations that alter the DNA sequence specificity of the catabolite gene activator protein of *E. coli*. *Nature* **311**, 232–235.

Ebright, R. H., Kolb, A., Buc, H., Kunkel, T. A., Krakow, J. S. & Beckwith, J. (1987). Role of glutamic acid-181 in DNA-sequence recognition by the catabolite gene activator protein (CAP) of *Escherichia coli*: altered DNA-sequence-recognition properties of [Val^{181}]CAP and [Leu^{181}]CAP. *Proc. natn. Acad. Sci. USA* **84**, 6083–6087.

Ebright, R. H., Le Grice, S. F. J., Miller, J. P. & Krakow, J. S. (1985). Analogs of cyclic AMP that elicit the biochemically defined conformational change in catabolite gene activator protein (CAP) but do not stimulate binding to DNA. *J. molec. Biol.* **182**, 92–107.

FERSHT, A. (1985). *Enzyme Structure and Mechanism*, 2nd ed. New York: W. H. Freeman & Co.

FILES, J. G. & WEBER, K. (1976). Limited proteolytic digestion of *lac* repressor by trypsin. *J. biol. Chem.* **251**, 3386–3391.

FRANKEL, A. D. & PABO, C. O. (1988). Fingering too many proteins. *Cell* **53**, 675.

FREDERICK, C. A., GRABLE, J., MELIA, M., SAMUDZI, C., JEN-JACOBSON, L., WANG, B.-C., GREENE, P. J., BOYER, H. W. & ROSENBERG, J. M. (1984). Kinked DNA in crystalline complex with EcoRI endonuclease. *Nature* **309**, 327–331.

FREEMONT, P. S., FRIEDMAN, J. M., BEESE, L. S., SANDERSON, M. R. & STEITZ, T. A. (1988). Cocrystal structure of an editing complex of Klenow fragment with DNA. *Proc. natn. Acad. Sci. USA* **85**, 8924–8928.

FREEMONT, P. S., OLLIS, D. L., STEITZ, T. A. & JOYCE, C. M. (1986). A domain of the Klenow fragment of *Escherichia coli* DNA polymerase I has polymerase but no exonuclease activity. *Proteins* **1**, 66–73.

FRIEDMAN, D. I. (1988). Integration host factor: a protein for all reasons. *Cell* **55**, 545–554.

GARGES, S. & ADHYA, S. (1985). Sites of allosteric shift in the structure of the cyclic AMP receptor protein. *Cell* **41**, 745–751.

GARTENBERG, M. R. & CROTHERS, D. M. (1988). DNA sequence determinants of CAP-induced bending and protein binding affinity. *Nature* **333**, 824–829.

GENT, M. E., GRONENBORN, A. M., DAVIES, T. W. & CLORE, G. M. (1987). *Biochem. J.* **242**, 645–653.

GEISLER, N. & WEBER, K. (1977). Isolation of the amino-terminal fragment of lactose repressor necessary for DNA binding *Biochemistry* **16**, 938–943.

GRINDLEY, N. D. F. (1983). Transposition of Tn3 and related transposons. *Cell* **32**, 3–5.

GRONENBORN, A. M. & CLORE, G. M. (1982). Proton nuclear magnetic resonance studies on cyclic nucleotide binding to the *Escherichia coli* adenosine cyclic 3′,5′-phosphate receptor protein. *Biochemistry* **21**, 4040–4048.

GRONENBORN, A. M., NERMUT, M. V., EASON, P. & CLORE, G. M. (1984). Visualization of cAMP receptor protein-induced DNA kinking by electron microscopy. *J. molec. Biol.* **179**, 751–575.

HATFULL, G. F. & GRINDLEY, N. D. F. (1988). *Genetic Recombination* (ed. G. Smith and R. Kucharlapati), pp. 357–564. Washington, D.C.: American Society for Microbiology.

HATFULL, G. F., SANDERSON, M. R., FREEMONT, P. S., RACCUIA, P. R., GRINDLEY, N. D. F. & STEITZ, T. A. (1989). Preparation of heavy atom derivatives using site-directed mutagenesis: introduction of cysteine residues into $\gamma\delta$ resolvase. *J. molec. Biol.* **208**, 661–667.

HECHT, M. H., NELSON, H. C. M. & SAUER, R. T. (1983). Mutations in λ repressor's amino-terminal domain: implications for protein stability and DNA binding. *Proc. natn. Acad. Sci. USA* **80**, 2676–2680.

HOCHSCHILD, A. & PTASHNE, M. (1986*a*). Cooperative binding of λ repressors to sites separated by integral turns of the DNA helix. *Cell* **44**, 681–687.

HOCHSCHILD, A. & PTASHNE, M. (1986*b*). Homologous interactions of λ repressor and λ *cro* with the λ operator. *Cell* **44**, 925–933.

HOL, W. G. S. (1985). The role of the α-helix dipole in protein function and structure. *Prog. Biophys. molec. Biol.* **45**, 149–195.

HOOPER, M. L., RUSSELL, R. L. & SMITH, J. D. (1972). Mischarging in mutant tyrosine transfer RNAs. *FEBS Lett.* **22**, 149.

IRWIN, N. & PTASHNE, M. (1987). Mutants of the catabolite activator protein of *Escherichia coli* that are specifically deficient in the gene-activation function. *Proc. natn. Acad. Sci. USA* **84**, 8315–8319.

JOHNSON, L. N. & PHILLIPS, D. C. (1965). Structure of some crystalline lysozyme-inhibitor complexes determined by X-ray analysis at 6 Å resolution. *Nature* **206**, 760–763.

JORDAN, R. S. & PABO, C. O. (1988). Structure of the λ complex at 2·5 Å resolution: details of the repressor–operator interactions. *Science* **242**, 893–899.

JORDAN, S. R., WHITCOMBE, T. V., BERG, J. M. & PABO, C. O. (1985). Systematic variation in DNA length yields highly ordered repressor–operator co-crystals. *Science* **230**, 1383.

JOYCE, C. M., OLLIS, D. L., RUSH, J., STEITZ, T. A., KONIGSBERG, W. H. & GRINDLEY, N. D. F. (1986). In: *Protein Structure, Folding and Design*, UCLA Symposia on Molecular and Cellular Biology (ed. D. Oxender), pp. 197–205. New York: Liss.

JOYCE, C. M. & STEITZ, T. A. (1987). DNA polymerase. I. From crystal structure to function via genetics. *Trends Biochem. Sci.* **12**, 288–292.

JURNAK, F. (1985). Structure of the GDP domain of EF-Tu and location of the amino acids homologous to *ras* oncogene proteins. *Science* **230**, 32–36.

KAPTAIN, R., ZUIDERWEG, E. R. P., SCHEEK, R. M., BOELENS, R. & VAN GUNSTERNE, W. F. (1985). A protein structure from nuclear magnetic resonance data. *J. molec. Biol.* **182**, 179–182.

KENNARD, O. & HUNTER, W. N. (1989). Oligonucleotide structure: a decade of results from single crystal X-ray diffraction studies. *Q. Reviews of Biophys.* **22**, 327–379.

KIM, R., MODRICH, P. & KIM, S.-H. (1984). 'Interactive' recognition in *Eco*R I restriction enzyme–DNA complex. *Nucl. Acids Res.* **12**, 7285–7292.

KIM, S. H., SUDDATH, F. L., QUIGLEY, G. J., MCPHERSON, A., SUSSMAN, J. L., WANG, A. H. J., SEEMAN, N. C. & RICH, A. (1974). Three-dimensional tertiary structure of yeast phenylalanine transfer RNA. *Science* **185**, 435–440.

KLENOW, H. & HENNINGSON, I. (1970). Selective elimination of the exonuclease activity of the DNA polymerase from *E. coli* B by a limited proteolysis. *Proc. natn. Acad. Sci. USA* **65**, 168.

KLUG, A., JACK, A., VISWAMITRA, M. A., KENNARD, O., SHAKKED, Z. & STEITZ, T. A. (1979). A hypothesis on a specific sequence-dependent conformation of DNA and its relation to the binding of the *lac*-repressor protein. *J. molec. Biol.* **131**, 669–680.

KOLB, A. & BUC, H. (1982). Is DNA unwound by the cyclic AMP receptor protein? *Nucl. Acids Res.* **10**, 473–485.

KOO, H.-S., WU, H.-M. & CROTHERS, D. M. (1986). DNA bending at adenine-thymine tracts. *Nature* **320**, 501–506.

KOUDELKA, G. B., HARRISON, S. C. & PTASHNE, M. (1987). Effect of non-contacted bases on the affinity of 434 operator for 434 repressor and *cro*. *Nature* **326**, 886–888.

KOUDELKA, G. B., HARBURY, P., HARRISON, S. C. & PTASHNE, M. (1988). DNA twisting and the affinity of bacteriophage 434 operator for bacteriophage 434 repressor. *Proc. natn. Acad. Sci. USA* **85**, 4633–4637.

KRAMER, H., NIEMOLLER, M., AMPUYAL, M., REVET, B., VON WILCKEN-BERGMANN, B. & MÜLLER-HILL, B. (1987). *lac* repressor forms loops with linear DNA carrying two suitably spaced *lac* operators *EMBO J.* **6**, 1481–1491.

LANDSCHULTZ, W. H., JOHNSON, P. R. & MCKNIGHT, S. L. (1988). The leucine zipper:

a hypothetical structure common to a new class of DNA binding proteins. *Science* **240**, 1759–1764.

Laughon, A. & Scott, M. P. (1984). Sequence of a *Drosophila* segmentation gene; protein structure homology with DNA-binding proteins. *Nature* **310**, 25–31.

Lawson, C. L., Zhang, R.-G., Schevitz, R. W., Otwinowski, Z., Joachimiak, A. & Sigler, P. B. (1988). Flexibility of the DNA-binding domains of *trp* repressor. *Proteins* **3**, 18–31.

Leahy, M. C. (1982). The binding of *lac* repressor to DNA substituted with nucleotide analogs. Ph.D. thesis, Yale University, New Haven, Connecticut.

Lee, M. S., Gippert, G. P., Soman, K. V., Case, D. A., Wright, P. E. (1989). Three-dimensional solution structure of a single zinc finger DNA-binding domain. *Science* **245**, 635–637.

Lehming, N., Sartorius, J., Niemöller, M., Genenger, G., Wilcken-Bergmann, B.v. & Müller-Hill, B. (1987). The interaction of the recognition helix of *lac* repressor with *lac* operator. *EMBO J.* **6**, 3145–3153.

Lewis, M., Wang, J. & Pabo, C. (1985). Structure of the operator binding domain of lambda repressor. In: *Biological Macromolecules and Assemblies*, vol. 2 (ed. F. A. Jurnak and A. McPherson), New York: John Wiley & Sons.

Liu-Johnson, H.-N., Gartenberg, M. R. & Grothers, D. M. (1986). The DNA binding domain and bending angle of *E. coli* CAP protein. *Cell* **47**, 995–1005.

Lomonossoff, G. P., Butler, P. J. G. & Klug, A. (1981). Sequence-dependent variation in the conformation of DNA. *J. molec. Biol.* **149**, 745–760.

Majors, J. (1977). Dissertation (Harvard University, Cambridge, MA).

McCall, M., Brown, T., Hunter, W. N. & Kennard, O. (1986). The crystal structure of d(GGATGGGAG) form an essential part of the binding site for TFIIIA. *Nature* **332**, 661–664.

Martin, K., Huo, L. & Schleif, R. F. (1986). The DNA loop model for ara repression: AraC protein occupies the proposed loop sites *in vivo* and repression-negative mutations lie in these same sites. *Proc. natn. Acad. Sci. USA* **83**, 3654–3658.

Matthews, B. W. (1988). No code for recognition. *Nature* **335**, 294–295.

Matthews, B. W., Ohlendorf, D. H., Anderson, W. F. & Takeda, Y. (1982). Structure of the DNA-binding region of *lac* repressor inferred from its homology with *cro* repressor. *Proc. natn. Acad. Sci. USA* **79**, 1428.

McClarin, J. A., Frederick, C. A., Wang, B.-C., Greene, P., Boyer, H. W., Grable, J. & Rosenberg, J. M. (1986). Structure of the DNA-*ECo*R I endonuclease recognition complex at 3 Å resolution. *Science* **234**, 1526–1541.

McKay, D. B., Pickover, C. A. & Steitz, T. A. (1982*a*). *E. coli lac* repressor is elongated with its DNA binding domains located at both ends. *J. molec. Biol.* **156**, 175–183.

McKay, D. B. & Steitz, T. A. (1981). Structure of catabolite gene activator protein at 2·9 Å resolution suggests binding to left-handed B-DNA. *Nature* **290**, 744–749.

McKay, D. B., Weber, I. T. & Steitz, T. A. (1982*b*). Structure of catabolite gene activator protein at 2·9 Å resolution: Incorporation of amino-acid sequence and interactions with c-AMP. *J. biol. Chem.* **257**, 9518–9524.

Miller, J. H. (1978). The *lacI* gene: its role in *lac* operon control and its use as a genetic system. In *The Operon* (ed. J. H. Miller and W. S. Reznikoff). Cold Spring Harbor, New York: Cold Spring Harbor Laboratory.

Miller, J., McLachlan, A. D. & Klug, A. (1985). Repetitive zinc-binding domains in the protein transcription factor IIIA from *Xenopus* oocytes. *EMBO J.* **4**, 1609–1614.

MOORE, S. (1981). *The Enzymes*, 3rd edn, vol. 14 (ed. P. D. Boyer), pp. 281–296. New York: Academic Press.

MORAS, D., COMARMOND, M. B., FISCHER, J., THEIRRY, J. C., EBEL, J. P. & GIEGÉ, R. (1980). Crystal structure of $tRNA^{Asp}$. *Nature* **288**, 669–674.

MÜLLER-HILL, B. (1975). *lac* repressor and *lac* operator. *Prog. Biophys. molec. Biol.* **30**, 227–252.

MÜLLER-HILL, B. (1983). Sequence homology between *lac* and *gal* repressors and three sugar-binding periplasmic proteins. *Nature* **302**, 163–164.

NELSON, H. C. M., FINCH, J. T., LUISI, B. F. & KLUG, A. (1987). The structure of an oligo(dA)·oligo(dT) tract and its biological implications. *Nature* **330**, 221–226.

NELSON, H. C. M. & SAUER, R. T. (1986). Interaction of mutant λ repressors with operator and non-operator DNA. *J. molec. Biol.* **192**, 22–38.

NORMANLY, J. & ABELSON, J. (1989). tRNA Identity. *Ann. Rev. Biochem.* **58**, 1029–1049.

OEFNER, C. & SUCK, D. (1986). Crystallographic refinement and structure of DNase I at 2 Å resolution. *J. molec. Biol.* **192**, 605–632.

OGATA, R. T. & GILBERT, W. (1979). DNA-binding site of *lac* repressor probed by dimethylsulfate methylation of *lac* operator. *J. molec. Biol.* **132**, 709–728.

OHLENDORF, D. H., ANDERSON, W. F., FISHER, R. G., TAKEDA, Y. & MATTHEWS, B. W. (1982). The molecular basis of DNA-protein recognition inferred from the structure of *cro* repressor. *Nature* **298**, 718–723.

OHLENDORF, D. H., ANDERSON, W. F., LEWIS, M., PABO, C. O. & MATTHEWS, B. W. (1983). Comparison of the structures of *cro* and λ repressor protein from bacteriophage λ. *J. molec. Biol.* **169**, 757–769.

OHLENDORF, D. H., ANDERSON, W. F., TAKEDA, Y. & MATTHEWS, B. W. (1983). High resolution structural studies of *cro* repressor protein and implications for DNA recognition. *J. biomol. Struct. Design* **1**, 553–563.

OHLENDORF, D. H. & MATTHEWS, B. W. (1983). Structural studies of protein-nucleic acid interactions. *Ann. Rev. Biophys. Bioeng.* **12**, 259–284.

OLLIS, D. L., BRICK, P., HAMLIN, R., XUONG, N. G. & STEITZ, T. A. (1985). Structure of large fragment of *Escherichia coli* DNA polymerase I complexed with dTMP. *Nature* **313**, 762–766.

O'SHEA, E. K., RUTTKOWSKI, R. & KIM, P. S. (1989). Evidence that the leucine zipper is a coiled coil. *Science* **243**, 538–542.

OTWINOWSKI, Z., SCHEVITZ, R. W., ZHANG, R.-G., LAWSON, C. L., JOACHIMIAK, A., MARMOSTEIN, R. Q., LUISI, B. F. & SIGLER, P. B. (1988). Crystal structure of *trp* repressor/operator complex at atomic resolution. *Nature* **335**, 321–329.

PABO, C. O. (1983). DNA–protein interactions. In *Proceedings of The Robert A. Welch Foundation Conferences on Chemical Research*, XXVII, *Stereospecificity in Chemistry and Biochemistry*, ch. 7, pp. 223–255. Houston, Texas.

PABO, C. O., KROVATIN, W., JEFFREY, A. & SAUER, R. T. (1982). The *N*-terminal arms of λ repressor wrap around the operator DNA. *Nature* **298**, 441–443.

PABO, C. O. & LEWIS, M. (1982). The operator-binding domain of λ repressor: structure and DNA recognition. *Nature* **298**, 443–447.

PABO, C. O. & SAUER, R. T. (1984). Protein–DNA recognition. *Ann. Rev. Biochem.* **53**, 293–321.

PABO, C. O., SAUER, R. T., STURTEVANT, J. M. & PTASHNE, M. (1979). THE λ repressor contains two domains. *Proc. natn. Acad. Sci. USA* **76**, 1608–1612.

PARRAGA, G., HORVATH, S. J., EISEN, A., TAYLOR, W. E., HOOD, L., YOUNG, E. T. &

Klevit, R. E. (1988). Zinc-dependent structure of a single-finger domain of yeast ADR1. *Science* **241**, 1489–1492.

Perona, J. J., Swanson, R. N., Rould, M. A., Steitz, T. A. & Söll, D. (1989). Structural basis for misaminoacylation by mutant *E. coli* glutaminyl-tRNA synthetase enzymes. *Science* **246**, 1152–1154.

Pflugrath, J. W. & Quiocho, F. A. (1985). Sulphate sequestered in the sulphate-binding protein of *Salmonella typhimurium* is bound solely by hydrogen bonds. *Nature* **314**, 257.

Phillips, S. E. V., Manfield, I., Parsons, I., Davidson, B. E., Rafferty, J. B., Somers, W. S., Margarita, D., Cohen, G. N., Saint-Girons, I. & Stockley, P. S. (1989). Cooperative tandem binding of *met* repressor of *Escherichia coli*. *Nature* **341**, 711–715.

Platt, T., Files, J. G. & Weber, K. (1973). *lac* repressor. *J. biol. Chem.* **248**, 110–121.

Porschke, D., Hillen, W. & Takahashi, M. (1984). The change of DNA structure by specific binding of the cAMP receptor protein from rotation diffusion and dichroism measurements. *EMBO J.* **3**, 2873–2878.

Price, P. A. (1975). The essential role of Ca^{2+} in the activity of bovine pancreatic deoxyribonuclease. *J. biol. Chem.* **250**, 1981–1986.

Ptashne, M. (1986). *A Genetic Switch*. Cambridge, MA: Cell Press.

Ptashne, M. (1986). Gene regulation by proteins acting nearby and at a distance. *Nature* **322**, 697–701.

Qian, Y. Q., Billeter, M., Otting, G., Müller, M., Gehring, W. J. & Wüthrich, K. (1989). The structure of the *Antennapedia* homeodomain determined by NMR spectroscopy in solution: Comparison with prokaryotic repressors. *Cell* **59**, 573–580.

Que, B. G., Downey, K. M. & So, A. (1978). Mechanism of selective inhibition of 3′ to 5′ exonuclease activity of *E. coli* DNA polymerase I by nucleoside 5′-monophosphates. *Biochemistry* **17**, 1603.

Rafferty, J. B., Somers, W. S., St.-Girons, I. & Phillips, S. E. V. (1989). Three-dimensional crystal structures of *E. coli met* repressor with and without corepressor. *Nature* **341**, 705–710.

Richmond, T. J., Finch, J. T., Rushton, B., Rhodes, D. & Klug, A. (1984). Structure of the nucleosome core particle at 7 Å resolution. *Nature* **311**, 532–537.

Richmond, T. J. & Steitz, T. A. (1976). Protein-DNA interaction investigated by binding *E. coli lac* repressor protein to poly[d(A·U-HgX)]. *J. molec. Biol.* **103**, 25–38.

Robertus, J. D., Ladner, J. E., Finch, J. T., Rhodes, D., Brown, R. S., Clark, B. F. C. & Klug, A. (1974). Structure of yeast phenylalanine tRNA at 3 Å resolution. *Nature* **250**, 546–551.

Rossman, M. G., Liljas, A., Branden, C.-I. & Banaszak, L. J. (1975). Evolutionary and structural relationships among dehydrogenases. In *The Enzymes*, vol II (ed. P. Boyer), pp. 61–102.

Rould, M. A., Perona, J. J., Söll, D. & Steitz, T. A. (1989). Structure of *E. coli* glutaminyl-tRNA synthetase complexed with $tRNA^{Gln}$ and ATP at 2·8 Å resolution: implications for tRNA discrimination. *Science* **246**, 1135–1142.

Rouvière-Yaniv, J. & Yaniv, M. (1979). *E. coli* DNA binding protein HU forms nucleosome-like structure with circular double-stranded DNA. *Cell* **17**, 265–274.

Satchwell, S. C., Drew, H. R., Travers, A. A. (1986). Sequence periodicities in chicken nucleosome core DNA. *J. molec. Biol.* **191**, 659–675.

SAUER, R. T., JORDAN, S. R., PABO, C. O. (1990). λ repressor: A model system for understanding protein-DNA interactions and protein stability. *Adv. Prot. Chem.* (in the press).

SAUER, R. T., PABO, C. O., MEYER, B. J., PTASHNE, M. & BACKMAN, K. C. (1979). Regulatory functions of the λ repressor reside in the amino-terminal domain. *Nature* **279**, 396–400.

SAUER, R. T., YOCUM, R. R., DOOLITTLE, R. F., LEWIS, M. & PABO, C. O. (1982). Homology among DNA-binding proteins suggests use of a conserved super-secondary structure. *Nature* **298**, 447–451.

SCHEFFLER, I. E., ELSON, E. L. & BALDWIN, R. L. (1968). Helix formation by dAT oligomers. I. Hairpin and straight-chain helices. *J. molec. Biol.* **36**, 291–304.

SCHEVITZ, R. W., OTWINOWSKI, Z., JOANCHIMIAK, A., LAWSON, C. L. & SIGLER, P. B. (1985). The three-dimensional structure of *trp* repressor. *Nature* **317**, 782–786.

SCHOLÜBBERS, H.-G., VAN KNIPPENBERG, P. H., BARANIAK, J., STEC, W. J., MORR, M. & JASTORFF, B. (1983). Investigations of stimulation of lac transcription *in vivo* in Escherichia coli by cAMP analogues. *Eur. J. Biochem.* **138**, 101–109.

SCHULMAN, L. H. & ABELSON, J. (1988). Recent excitement in understanding transfer RNA identity. *Science* **240**, 1591–1592.

SCHULMAN, L. H. & PELKA, H. (1985). *In vitro* conversion of a methionine to a glutamine-acceptor tRNA. *Biochemistry* **24**, 7309–7314.

SCHULTZ, S. C., SHIELDS, G. C. & STEITZ, T. A. (1990). Crystallization of *E. coli* CAP with its operator DNA: the use of modular DNA. *J. molec. Biol* (in the press).

SEEMAN, N. C., ROSENBERG, J. M., RICH, A. (1976). Sequence-specific recognition of double helical nucleic acids by proteins. *Proc. natn. Acad. Sci. USA* **73**, 804–808.

SEONG, B. L., LEE, C.-P. & RAJBHANDARY, U. L. (1989). Supression of amber codons *in vivo* as evidence that mutants derived fro *Escherichia coli* initiator tRNA can act at the step of elongation in protein synthesis. *J. biol. Chem.* **264**, 6504.

SHEPHERD, J. C. W., MCGINNIS, W., CARRASCO, A. E., DE ROBERTS, E. M., & GEHRING, W. J. (1984). Fly and frog homoeo domains show homologies with yeast mating type regulatory proteins. *Nature* **310**, 5972, 70–71.

SHIMURA, Y., AONO, H., OZEKI, H., SARABHAI, A., LAMFORM, H. & ABELSON, J. (1972). Mutant tyrosine tRNA of altered amino acid specificity. *FEBS Lett.* **22**, 144–148.

SIMPSON, R. B. (1980). Interaction of the cAMP receptor protein with the *lac* promoter. *Nucl. Acids Res.* **8**, 759.

STEITZ, T. A., BEESE, L., FREEMONT, P. S., FRIEDMAN, J. & SANDERSON, M. R. (1987). Structural studies of Klenow fragment: an enzyme with two active sites. *Cold Spring Harbor Symposia on Quantitative Biology*, ch. 52, pp. 465–471. Cold Spring Harbor, NY: Cold Spring Harbor Laboratory.

STEITZ, T. A. OHLENDORF, D. H., MCKAY, D. B., ANDERSON, W. F. & MATTHEWS, B. W. (1982). Structural similarity in the DNA binding domains of catabolite gene activator and cro repressor proteins. *Proc. natn. Acad. Sci. USA* **79**, 3097–3100.

STEITZ, T. A., RICHMOND, T. J., WISE, D. & ENGELMAN, D. M. (1974). The *lac* repressor protein: molecular shape, subunit structure and proposed model for operator interaction based on structural studies of micro-crystals. *Proc. natn. Acad. Sci. USA* **72**, 53.

STEITZ, T. A., STENKAMP, R. E., GEISLER, N., WEBER, K. & FINCH, J. (1979). X-ray and electron microscopic studies of crystals of core *lac* repressor protein. In *Biomolecular Structure, Conformation, Function and Evolution* (ed. R. Srinivasan)., Oxford: Pergamon Press.

STEITZ, T. A. & WEBER, I. T. (1985). Structure of catabolite gene activator protein. In *Biological Macromolecules and Assemblies*, 2nd edn (ed. A. McPherson and F. Jurnak), pp. 290–321. New York: John Wiley.

STEITZ, T. A., WEBER, I. T., OLLIS, D. & BRICK, P. (1983). Crystallographic studies of protein–nucleic acid interaction: catabolite gene activator protein and the large fragment of DNA polymerase I. *J. biomolec. Struct. Dyn.* **1**, 1023–1037.

SUCK, D., LAHM, A. & OEFNER, C. (1988). Structure refined to 2 Å of a nickel DNA octanucleotide complex with DNase I. *Nature* **332**, 6163, 465–468.

SUCK, D. & OEFNER, C.(1986). Structure of DNase I at 2·0 Å resolution suggests a mechanism for binding to and cutting DNA. *Nature* **321**, 620–625.

SUCK, D., OEFNER, C. & KABSCH, W. (1984). Three-dimensional structure of bovine pancreatic DNase I at 2·5 Å resolution. *EMBO J.* **3**, 2423–2430.

SUNG, M. T. & DIXON, G. H. (1970). Modification of histones during spermiogenesis in trout: a molecular mechanism of altering histone binding to DNA. *Proc. natn. Acad. Sci. USA* **67**, 1616–1623.

TAKEDA, Y., OHLENDORF, D. H., ANDERSON, W. F. & MATTHEWS, B. W. (1983). DNA-binding proteins. *Science* **221**, 1020–1026.

TANAKA, I., APPELT, K., DIJ, K. L., WHITE, S. W. & WILSON, K. S. (1984). 3 Å resolution structure of a protein with histone-like properties in prokaryotes. *Nature* **310**, 376–381.

VYAS, N. K., VYAS, M. N., & QUIOCHO, F. A. (1988). Sugar and signal-transducer binding sites of the *Escherichia coli* galactose chemoreceptor protein. *Science* **242**, 1290–1295.

WANG, B.-C. (1987). Resolution of phase ambiguity in macromolecular crystallography. *Methods in Enzymol.* **115**, 90–111.

WARRANT, R. W. & KIM, S. -H. (1978). α-Helix–double helix interaction shown in the structure of a protamine-transfer RNA complex and a nucleoprotamine model. *Nature* **271**, 130–135.

WARWICKER, J., ENGELMAN, B. P. & STEITZ, T. A. (1987). Electrostatic calculations and model building suggest that DNA bound to CAP is sharply bent. *Proteins* **2**, 283–289.

WEBER, K. & FILES, J. G. (1976). Limited proteolytic digestion of *lac* repressor by trypsin. *J. biol. Chem.* **251**, 3386–3391.

WEBER, I. T. & STEITZ, T. Q. (1984). A model for non-specific binding of catabolite gene activator protein to DNA. *Nucl. Acids Res.* **12**, 8475–8487.

WEBER, I. T. & STEITZ, T. A. (1984). Model of specific complex between CAP and B-DNA suggested by electrostatic complementarity. *Proc. natn. Acad. Sci. USA* **81**, 3973–3977.

WEBER, I. T. & STEITZ, T. A. (1987). The structure of a complex of catabolite gene activator protein and cyclic AMP refined at 2·5 Å resolution. *J. molec. Biol.* **198**, 311–326.

WEBER, I. T., MCKAY, D. B. & STEITZ, T. A. (1982*a*). Two helix DNA binding motif of CAP found in *lac* repressor and *gal* repressor *Nucl. Acids Res.* **10**, 5085–5102.

WEBER, I. T., STEITZ, T. A., BUBIS, J. & TAYLOR, S. S. (1987). Predicted structures of cAMP binding domains of type I and II regulatory subunits of cAMP-dependent protein kinase. *Biochemistry* **26**, 343–351.

WEBER, I. T., TAKIO, K., TITANI, K. & STEITZ, T. A. (1982*b*). The cAMP-binding domains of the regulatory subunit of cAMP-dependent protein kinase and the catabolite gene activator proton are homologous. *Proc. natn. Acad. Sci. USA* **79**, 7679–7683.

WEBER, P. C., OLLIS, D. L., DEBRIN, W. R., ABDEL-MEGUID, S. S. & STEITZ, T. A. (1982*c*). Crystallization of resolvase, a repressor which also catalyzes site-specific DNA recombination. *J. biol. Chem.* **157**, 689–690.

WHARTON, R. (1985). Thesis, Harvard University, Cambridge, MA.

WHARTON, R. P. & PTASHNE, M. (1985). Changing the binding specificity of a repressor by redesigning an α-helix. *Nature* **316**, 601–605.

WHARTON, R. P., BROWN, E. L. & PTASHNE, M. (1984). Substituting an α-helix switches the sequence-specific DNA interactions of a repressor. *Cell* **38**, 361–369.

WOLBERGER, C., DONG, Y., PTASHNE, M. & HARRISON, S. C. (1988). Structure of a phage 434 cro/DNA complex. *Nature* **335**, 789–795.

WOO, N. H., ROE, B. A. & RICH, A. (1980). Three-dimensional structure of *Escherichia coli* initiator tRNAf*Met*. *Nature* **286**, 346–351.

WOODBURY, C. P., HAGENBÜCHLE, O. & VON HIPPEL, P. H. (1980). DNA site recognition and reduced specificity of the *Eco*r I endonuclease. *J. biol. Chem.* **255**, 11534–11546.

WU, H. & CROTHERS, D. M. (1984). The locus of sequence-directed and protein-induced DNA bending. *Nature* **308**, 509–513.

YANG, C.-C. & NASH, H. W. (1989). The interaction of *E. coli* IHF protein with its specific-binding sites. *Cell* **57**, 869–880.

YANIV, M., FOLK, W., BERG, P. & SOLL, L. (1974). A single mutational modification of a tryptophan-specific transfer RNA permits aminoacylation by glutamine and translation of the codon UAG. *J. molec. Biol.* **86**, 245–260.

YARUS, M. (1988). tRNA identity: a hair of the dogma that bit us. *Cell* **55**, 739–741.

YARUS, M., KNOWLTON, R. & SOLL, L. (1977). Aminoacylation of the ambivalent Su+7 amber suppressor tRNA. In *Nucleic Acid–Protein Recognition* (ed. H. J. Vogel) pp. 391–409. New York: Academic Press.

YOON, C., PRIVE, G. G., GOODSELL, D. S. & DICKERSON, R. E. (1988). Structure of an alternating-B DNA helix and its relationship to A-tract DNA. *Proc. natn. Acad. Sci. USA* **85**, 6332–6336.

YOUNG, T.-S., KIM, S.-H., MODRICH, P., SETH, A. & JAY, E. (1981). Preliminary X-ray diffraction studies of *Eco*R I restriction endonuclease–DNA complex. *J. molec. Biol.* **145**, 607–610.

ZELWER, C., RISLER, J. L. & BRUNIE, S. (1982). Crystal structure of *Escherichia coli methionyl-tRNA synthetase at 2·5 Å resolution. J. molec. Biol.* **115**, 63–81.

ZHANG, R.-G., JOACHIMIAK, A., LAWSON, C. L., SCHEVITZ, R. W., OTWINOWSKI, Z. & SIGLER, P. G. (1987). The crystal structure of *trp* aporepressor at 1·8 Å shows how binding tryptophan enhances DNA affinity. *Nature* **327**, 591–597.

ZUBAY, G. & DOTY, P. J. (1959). The isolation and properties of deoxyribonucleoprotein particles containing single nucleic acid molecules. *J. molec. Biol.* **7**, 1–20.

"Recently J. Rosenberg and colleagues have reinterpreted the structure of *Eco*RI in light of an improved experimental electron density map. Due to earlier errors in tracing the polypeptide backbone, the amino acid sequence numbers given in Figs. 30 and 31 and in the accompanying text are not correct. The general conclusions, however, are largely supported in the new interpretation."

Index